KU-714-536

Mathematics

Michael Jennings

Contents

Examination boards

AQA Assessment and Qualifications Alliance
Devas Street,
Manchester,
M15 6EX
www.aqa.org.uk

EDEXCEL
Stewart House,
32 Russell Square,
London,
WC1B 5DN
www.edexcel.org.uk

OCR Oxford Cambridge and RSA Examinations
1 Hills Road,
Cambridge,
CB1 2EU
www.ocr.org.uk

CCEA Northern Ireland Council for Curriculum, Examinations and Assessment
29 Clarendon Road,
Belfast,
BT1 3BG
www.ccea.org.uk

WJEC Welsh Joint Education Committee
245 Western Avenue,
Cardiff,
CF5 2YX
www.wjec.co.uk

This book is based on a new specification for AS Mathematics which will be taught in schools and colleges from September 2004 for first examination in June 2005. All five examination boards (**Edexcel**, **AQA**, **OCR**, **CCEA**, and **WJEC**) have produced their own specification but the general structure of all of these specifications is broadly the same.

To obtain an AS level qualification in Mathematics, candidates must study:
two (Pure) Core units, C1 and C2, plus one Applied unit.

The Applied unit could be in Mechanics (M1), Statistics (S1) or in Decision or Discrete Mathematics (D1).

Candidates are not permitted the use of a calculator in the C1 unit but can use a calculator (scientific or graphical) in the other units.

Each unit will be assessed via a one and half hour written paper.

There is no coursework, except for AQA, who have coursework versions of their Applied units.

N.B. The above is a general outline and there are one or two exceptions/variations.

More detail on each of the new AS mathematics specifications is available at the appropriate website:

www.edexcel.org.uk www.aqa.org.uk
www.ocr.org.uk www.ccea.org.uk
www.wjec.org.uk

AS exams

Different types of questions

Questions on mathematics papers have varying degrees of structure to them. Those at the beginning of a paper tend to be short and sharp and worth only a few marks and thus there is little opportunity to break the question down into smaller parts. However, as you progress through the paper, the questions become longer and more challenging and will often have several parts to them. These parts could be totally independent or there may be a common theme running through a question where the examiner is attempting to lead you through the early parts in order to give you a hint on a method that might be used to do a later, more difficult part – the wording "hence, or otherwise" is an indication of this type of question.

A question can also be broken down into smaller parts in order for a candidate who gets stuck on an early part to, nonetheless, be able to go on and score marks on later parts. To this end, the answer may be given in a particular part and the candidates be asked to show it. Thus if a candidate is unable to derive the correct result, he or she can use the printed answer to hopefully progress further through the question.

Sometimes candidates may be required, as the first part of a question, to produce a proof or derivation of a standard result or formula which is then used in a subsequent part or parts to solve a particular problem. Explanations or definitions of particular terms may also be asked for, particularly on statistics papers.

You may also be asked to comment on or interpret a result (in statistics or decision mathematics), explain a modelling assumption and where it is used (in mechanics), identify a flaw in an argument (in pure mathematics) or how a particular model could be refined to make it more realistic.

What examiners look for

- clear and concise methods, although any valid method, no matter how long, will be given full credit.
- appropriate and accurate use of notation and symbolism.
- large and clearly labelled diagrams and graphs where appropriate.
- appropriate and accurate use of technology (e.g. a calculator).
- the ability to interpret and comment on results obtained.

What makes an A, C and E grade candidate?

- **A grade candidates** have a broad knowledge of mathematics and can apply that knowledge in a wide variety of situations, including unfamiliar scenarios, accurately and efficiently. They are strong on all of the units.The minimum mark for a grade A is 80% on the Uniform Mark Scale.
- **C grade candidates** have a fair knowledge of mathematics but find it less easy to apply their knowledge in unfamiliar situations. Their work is less accurate and they have weaknesses on some of the units. The minimum mark for a grade C is 60% on the Uniform Mark Scale.
- **E grade candidates** have a poor knowledge of mathematics and are unable to apply it in unfamiliar situations. Their work has many errors and they are unable to recall key facts and techniques. The minimum mark for a grade E is 40% on the Uniform Mark Scale.

Successful revision

Revision skills

- By far the best way to revise for mathematics is by doing mathematics i.e. by solving problems. Of course you have to learn the theory but unless you can apply the theory to actually tackle problems, the knowledge is of little use. It is therefore essential, if your preparation is to be effective, that you encounter as many different situations and scenarios as possible, by doing as many practice questions as possible, so that you can learn how to recognise which techniques will be appropriate to solve a particular type of problem and which will not.
- When revising the theory, try to summarise, as concisely as possible, the key points and how they relate to other parts of the syllabus. Writing out your own concise notes (the briefer the better) for each syllabus topic can be a good way of learning material.

Practice questions

To use this book effectively

- Examine the grade A and grade C sample answers and make sure that you understand where the errors have been made and how to correct them.
- Try the exam practice questions – don't be tempted to look at the answers too quickly if you get stuck; you will learn a great deal more from a question if you struggle with it and eventually sort it out or at least make some progress, by yourself, using worked examples in your notes or in a textbook to guide you.
- When you feel confident and ready, try the mock exam papers.

Common errors

Many errors occur due to careless work with signs, particularly when removing brackets, and errors in basic algebra and trigonometry.

1. **Many of the most common errors occur as a result of students treating all functions, f, as being linear**

   i.e. $f(a + b) = f(a) + f(b)$, for all a and b.

   e.g. $(a + b)^2 = a^2 + b^2$, or similar

   $\sqrt{(a + b)} = \sqrt{a} + \sqrt{b}$

   $\frac{1}{a + b} = \frac{1}{a} + \frac{2}{b}$

   $\sin (A + B) = \sin A + \sin B$

   $\ln (A + B) = \ln A + \ln B$

   $e^{x + y} = e^x + e^y$

   Of course none of the above are true, **in general** (some of them may be true in certain special cases). See if you can, where possible, correct them.

2. **Confusion with notation**

   e.g.

   f^{-1}, the **inverse** of f, is often confused with f', the **derivative** of f

   fg means "do g first then f", not the other way round.

How to boost your grade

- Ensure that you do exactly as the question says, e.g. if you are told to use a particular method then you will receive no credit whatsoever for using a different method, even if you get the question right.
- Ensure that you give answers to the correct degree of accuracy when requested to do so – you will definitely lose marks if you don't.
- Show your working – a very high proportion of the available marks at A level are Method Marks.
- You can answer the questions in any order that you like – you should attempt a few of the shorter questions at the beginning of the paper to boost your confidence, making sure that you leave yourself plenty of time for the last two or three questions, for which there are a very high number of marks.
- For the shorter questions, make life easier for the examiner, by ruling off at the end of a question and either leave a space before you start the next question, or if you are near the bottom of the page, start on a fresh piece of paper. Always start the longer questions on a fresh page. This will help to avoid copying and transcription errors which are made when turning over a page.
- If you attempt a question using two different methods, then do not cross either of them out but instead leave both – the examiner will mark both and award you the better mark.
- Dimensional analysis, particularly in algebraic Mechanics questions, will often help to spot silly mistakes. For example, if you are asked to find the loss in kinetic energy in a particular problem and you obtain an answer of $5mu$, you should realise that you have made an error as this expression has momentum (or impulse) units.
- Don't work too quickly – try to check each line of working before moving on to the next – but on the other hand don't waste time e.g. by underlining everything; use your time sensibly – try to match the time that you use to the marks available for that question – if you get stuck on a question, particularly a short one, don't panic! Leave a space and go back to it later, if you have time.
- Familiarise yourself with the Formulae Booklet before you do the exam – make sure that you know what is in there and where it is situated.
- Make sure that you have a calculator with you when permitted (you are not allowed calculators for certain pure units), and that it works!
- Put all your past papers together and look through them so that you are familiar with the type of questions that are asked and look through your copy of the syllabus to make sure that you have revised all the topics.

Glossary of terms used in examination questions

Prove – Show that a result is true, using a reasoned argument which starts from accepted basic results (the question will sometimes clarify what you can assume).

Write down, state – no justification is needed for your answer.

Calculate, find, determine, show, solve – show sufficient working to make your method clear. (N.B. Answers without working will gain no credit).

Deduce, hence – use the given result or previous part to establish the result.

Sketch – graph paper not needed; show the general shape of a graph, where it crosses the axes (if it does), any asymptotes and any points of particular significance.

Draw – plot accurately on graph paper using a suitable scale.

Find the exact value – leave your answer as a fraction or in surds, or in terms of logarithms, exponentials or π; note that using a calculator is likely to introduce decimal approximations, resulting in a loss of marks.

Chapter 1

Core 1

Questions with model answers

C grade candidate – mark scored 7/13

N.B. Calculators cannot be used for C1 questions

(1) The straight line *l* passes through the point (−2, 1) and is perpendicular to the line *m* with equation $2y - x + 11 = 0$.

(a) Find an equation for *l*. [6]

Equation of m is $y = \frac{1}{2}x - \frac{11}{2} \rightarrow$ gradient $= \frac{1}{2}$

$\rightarrow$ gradient of l is 2

Equation of l is $y - 1 = 2(x - -2)$

$\rightarrow y = 2x + 5.$

(b) Find the coordinates of the point where the lines *l* and *m* intersect. [4]

$y = \frac{1}{2}x - \frac{11}{2}$ and $y = 2x + 5$

i.e. $\frac{1}{2}x - \frac{11}{2} = 2x + 5$

$\rightarrow \frac{3x}{2} = \frac{21}{2}$

$\rightarrow x = 7 \rightarrow y = 19.$

(c) Verify that the point *A*(5, −3) lies on the line *m*. [1]

$2 \times (-3) - 5 + 11 = 0.$

(d) Deduce the perpendicular distance of A from the line l. [2]

l
X
A(5, −3)
m
(−2, 1)

Examiner's Commentary

For help see Revise AS Study Guide section 1

Correct; has put in form $y = mx + c$ and read off the value of m.
Has correctly inverted but forgotten to change the sign to obtain the gradient of the normal.
Correct use of $y - y_1 = m(x - x_1)$ but wrong value of m.
Incorrect answer, **4/6 scored**.

Correct substitution method.
Another sign error!

Incorrect, **2/4 scored**.

An easy mark, 1 scored.

No answer given, **0 scored**.
The idea is that since the lines intersect at right angles at *X*, say, the perpendicular distance of *A* from l is the length of *AX*, where *X* should be (−7, −9) (see diagram).

GRADE BOOSTER

Check each line of your working carefully before writing down the next line to avoid cumulative errors.

A grade candidate – mark scored 9/11

Examiner's Commentary

(2) (a) Find the values of the constants a, b and c such that $a(x + b)^2 + c = 4x^2 - 24x + 27$, for all x. [4]

$a(x^2 + 2bx + b^2) + c = 4x^2 - 24x + 27$

$a = 4;\ 8b = -24,\ \ b = -3;$

$4b^2 + c = 27,\ \ 36 + c = 27,\ \ c = -9.$

Brackets correctly expanded.

Coefficients of x^2 and x equated and the constants found. 4/4 scored.

(b) Hence, or otherwise, find the set of values of x for which $4x^2 - 24x + 27 \geqslant 0$. [3]

$4(x - 3)^2 - 9 \geqslant 0,$

$x - 3 \geqslant \frac{3}{2}$ or $-\frac{3}{2}$

$x \geqslant \frac{9}{2}$ or $\frac{3}{2}.$

This is an easy mistake to make; particular care needs to be taken when square rooting inequalities. Should be: $x \geqslant \frac{9}{2}$ or $x \leqslant \frac{3}{2}$, 2/3 scored.

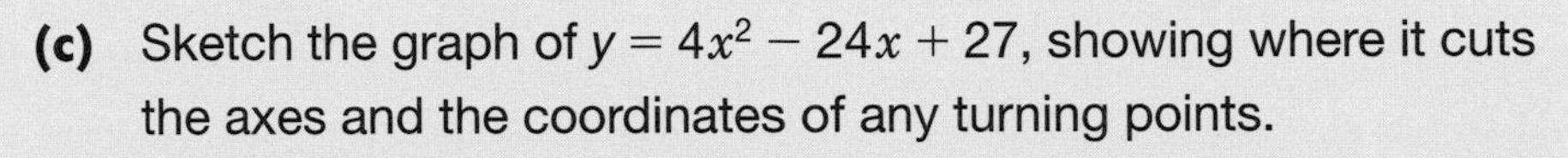

(c) Sketch the graph of $y = 4x^2 - 24x + 27$, showing where it cuts the axes and the coordinates of any turning points. [4]

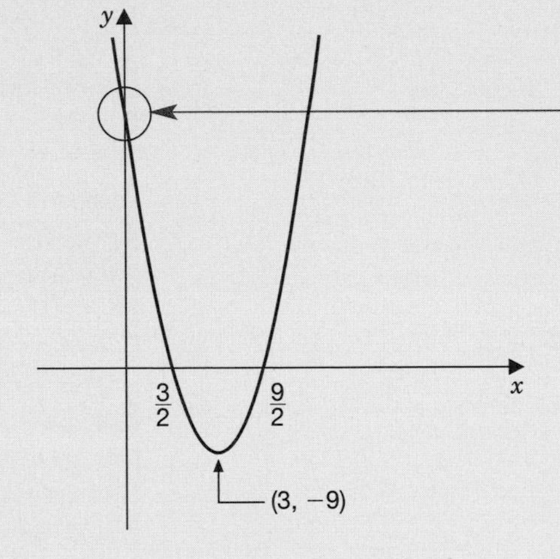

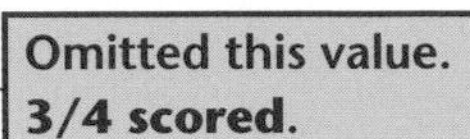

Omitted this value. 3/4 scored.

For help see Revise AS Study Guide section 1

Core 1

Exam practice questions

1.1 Algebra and functions

1 The roots of a quadratic equation $x^2 + bx + c = 0$ are $2+\sqrt{5}$ and $2-\sqrt{5}$. Find the values of b and c. [5]

2 Solve the simultaneous equations $2x - y = -1$
$xy + y = 3$. [5]

3 **(a)** Given that $y = 2^x$, show that $2^{2x+2} = 4y^2$. [2]

(b) Hence, or otherwise, solve the equation $2^{2x+2} + 7(2)^x - 2 = 0$. [5]

4 Simplify the following, giving your answer in the form $a + b\sqrt{2}$, where a and b are rational numbers,

$$\frac{2}{(\sqrt{2} - 1)} + \frac{3}{(\sqrt{2} + 1)}$$ [5]

5 **(a)** Find the values of the constants a, b and c such that

$$2x^2 + 8x - 1 = a(x + b)^2 + c \quad \text{for all } x.$$ [5]

(b) Hence, or otherwise, solve the equation $2x^2 + 8x - 1 = 0$, giving your answers in the form $a + b\sqrt{2}$. [4]

(c) Sketch the graph with equation $y = 2x^2 + 8x - 1$, giving the coordinates of any points where the graph crosses the coordinate axes, and any stationary points. [3]

(d) Use your graph to solve

(i) $2x^2 + 8x - 1 = -9$, [2]

(ii) $2x^2 < 1 - 8x$. [2]

6 A farmer wishes to use 160 m of fencing to construct three equal square enclosures, each of side x m, together with one larger square enclosure, of side y m.

(a) Write down an equation in x and y. [2]

The total area of the enclosures is to be 508 m^2.

(b) Write down a further equation in x and y. [2]

(c) Hence determine the dimensions of the enclosures. [6]

7 **(a)** Simplify $(x^{\frac{1}{2}} + x^{-\frac{1}{2}})(x^{\frac{1}{2}} - x^{-\frac{1}{2}})$, giving your answer over a common denominator. [4]

(b) Hence, or otherwise, write $\frac{8}{3}$ in the form $(a + b)(a - b)$ where a and b are numbers. [3]

Answers on pages 16–26 Answers on pages 16–26 Answers on pages 16–26

8 **(a)** Find the values of the constants p, q and r such that

$$6x - 5 - x^2 \equiv p(x + q)^2 + r.$$ [5]

(b) Deduce the maximum value of $6x - 5 - x^2$, stating the value of x for which it occurs. [2]

(c) Sketch the graph of $y = 6x - 5 - x^2$, giving the coordinates of any points where it crosses the coordinate axes and also any turning points. [5]

9 Given that the equation

$$x^2 - (k + 1)x + 1 = 0, \text{ where } k \text{ is a constant,}$$

has two real distinct solutions,

(a) write down an inequality which must be satisfied by k [2]

(b) find the range of possible values of k. [4]

10 **(a)** Solve the equation

$$2x^2 - 4x + 1 = 0,$$

expressing your answers in the form $a + b\sqrt{2}$, where a and b are rational numbers. [4]

(b) Find, simplifying your answers in each case,

(i) the sum of the roots of the equation $2x^2 - 4x + 1 = 0$, [2]

(ii) the product of the roots of the equation $2x^2 - 4x + 1 = 0$. [3]

11 The figure shows a sketch of the curve with equation $y = f(x)$. On separate diagrams, sketch the curve with equation

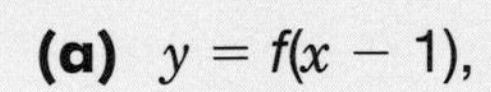

(a) $y = f(x - 1)$, [4]

(b) $y = f(\frac{1}{2}x)$. [5]

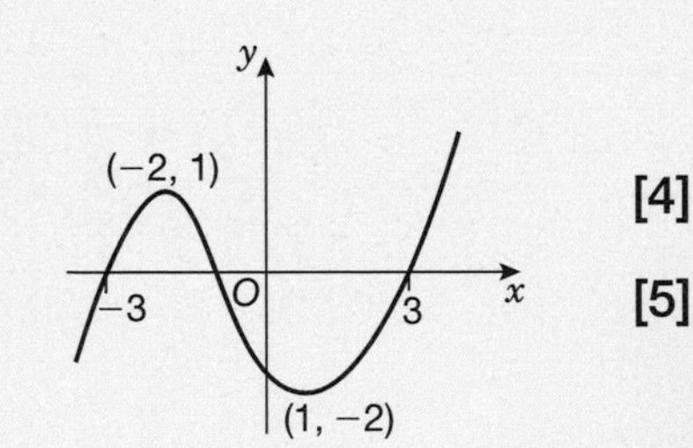

On each diagram, show clearly the coordinates of the minimum point, and of each point at which the curve crosses the coordinate axes.

1.2 Coordinate geometry

1 The point A has coordinates $(-4, 2)$ and the point B has coordinates $(-2, 4)$.

(a) Find an equation of the straight line l which passes through the origin O and the mid-point of AB. [4]

(b) Find an equation of the straight line m which passes through B and the mid-point of OA. [2]

(c) Find the point of intersection of the lines l and m. [2]

Answers on pages 16–26 Answers on pages 16–26 Answers on pages 16–26

2 The graph shows part of the curve with equation $y = 16x - kx^2$, where k is a constant. The points A and D have coordinates (0, 18) and (6, 15) respectively.

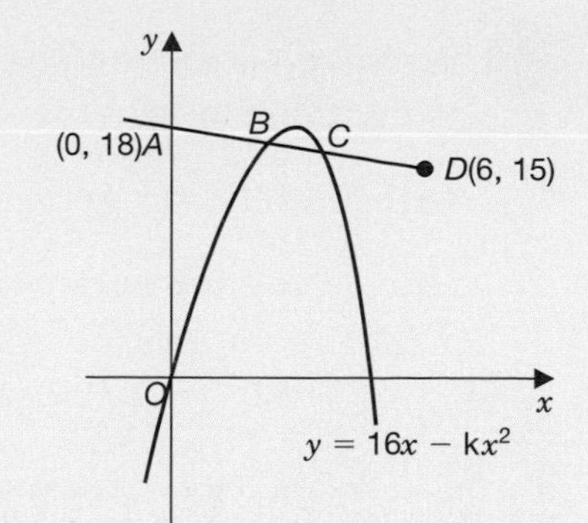

(a) Calculate, giving your answer to 3 significant figures, the length of AD. [2]

The line l passes through the points A and D and intersects the curve at the points B and C, as shown.

(b) Obtain an equation of l in the form $y = mx + c$, where m and c are constants. [4]

Given also that C has coordinates (4, 16),

(c) show that $k = 3$, [2]

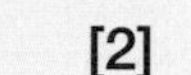

(d) calculate the coordinates of B. [4]

3 Find the equation of the straight line passing through the point $(-1, 3)$ which is parallel to the straight line with equation $2x + 3y - 4 = 0$, giving your answer in the form $ax + by + c = 0$, where a, b and c are integers. [6]

Core 1

4 The point A has coordinates $(-2, 5)$, the point B has coordinates (4, 3) and M is the mid-point of AB.

(a) Find the coordinates of M. [2]

(b) Find the equation of the perpendicular bisector of the line AB. [4]

(c) Verify that the point $P(2, 7)$ is on this perpendicular bisector. [1]

(d) Deduce the perpendicular distance of P from the line AB. [3]

5 **(a)** Find the gradient of the line L with equation $3x + 2y + 3 = 0$. [2]

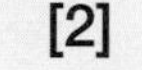

(b) Hence find, giving your answer to the nearest degree, the angle that L makes with the positive x-axis. [3]

6 The straight line L_1 has equation $2x + 4y + 1 = 0$ and the straight line L_2 has equation $2y = 3 - x$.

(a) Show that L_1 and L_2 are parallel. [2]

(b) Verify that the point $A(-2.5, 1)$ lies on L_1 and the point $B(-1.8, 2.4)$ lies on L_2. [2]

(c) Show that AB is perpendicular to L_1. [2]

(d) Deduce the perpendicular distance between L_1 and L_2, giving your answer as a multiple of $\sqrt{5}$. [5]

Answers on pages 16–26 **Answers** on pages 16–26 **Answers** on pages 16–26

7 A straight line L is to be drawn through the point $(-1, 3)$ and perpendicular to the line M with equation $3x - 4y - 10 = 0$.

(a) Find the equation of L. [5]

(b) Find the coordinates of the point where the line L meets the line M. [4]

1.3 Sequences and series

1 An arithmetic series has third term -20 and eleventh term 20.

(a) Find the first term and the common difference. [5]

The sum of the first k terms of the series is zero.

(b) Find the non-zero value of k. [4]

2 Mark is given an interest-free loan to buy a second-hand car. He repays the loan in monthly instalments. He repays £20 the first month, £22 the second month and the repayments continue to rise by £2 per month until the loan is repaid. Given that the final monthly repayment is £114,

(a) show that the number of months that it will take Mark to repay the loan is 48, [3]

(b) find, in pounds, the price of the car. [4]

3 A sequence of numbers $u_1, u_2, \ldots, u_n, \ldots$ is given by the formula $u_n = 3\left(\frac{2}{3}\right)^n - 1$, where n is a positive integer.

(a) Find the values of u_1, u_2, and u_3. [2]

(b) Find $\sum_{n=1}^{15} 3\left(\frac{2}{3}\right)^n$, and hence show that $\sum_{n=1}^{15} u_n = -9.014$ to 4 significant figures. [5]

(c) Prove that $3u_{n+1} = 2u_n - 1$. [4]

4 **(a)** Calculate:

(i) $\sum_{r=1}^{50} (3r - 7)$ **(ii)** $\sum_{r=1}^{30} (3r - 7)$. [5]

(b) Hence, or otherwise, find the value of $\sum_{r=31}^{50} (3r - 7)$. [2]

5 The sum, S_n, of the first n terms of an arithmetic series is given by

$S_n = an + bn^2$, where a and b are constants.

Given that the sum of the first three terms of the series is 18 and that the sum of the first five terms of the series is 50,

(a) find the values of a and b. [5]

(b) Deduce, or find otherwise, an expression, in terms of n, for the nth term of the arithmetic series. [4]

(c) Hence find the common difference of the arithmetic series. [2]

Answers on pages 16–26 **Answers** on pages 16–26 **Answers** on pages 16–26

6 The sum of the first n terms of an arithmetic series is given by $\frac{1}{2}n(3n + 5)$.

By giving n suitable values, find:

(a) the first term, [1]

(b) the common difference, [2]

(c) the sum of the first 20 terms. [3]

7 The first and last terms of an arithmetic progression are -2 and 73 respectively, and the sum of all the terms is 923. Find:

(a) the number of terms, [5]

(b) the common difference. [4]

8 The first term of an arithmetic progression is 2 and the common difference of the progression is 3. Find:

(a) the sum to n terms of this progression, [2]

(b) the value of n for which the sum of this progression is 610. [3]

9 A man borrows £10 000 from a bank and repays it by 36 monthly payments as follows. His first payment is £100 and he then increases his monthly payments by £20 every month. Find:

(a) his final payment, [2]

(b) the total interest paid on the loan. [4]

Core 1

1.4 Differentiation

1 A curve has equation $y = 2x - x^2$

(a) Find the equation of the normal to the curve at the origin O. [5]

(b) Find the coordinates of the point where the normal cuts the curve again. [5]

2 Given that $y = (x^{\frac{1}{2}} + x^{\frac{3}{2}})^2$, find

(a) $\frac{dy}{dx}$, [4]

(b) the value of x for which $\frac{d^2y}{dx^2}$ equals zero. [2]

3 A certain curve has equation $y = ax^2 + bx + c$ and passes through the point P with coordinates $(2, -1)$. Given that, at the point P, $\frac{dy}{dx} = 3$ and $\frac{d^2y}{dx^2} = -2$, find the values of a, b and c. [6]

Answers on pages 16–26 Answers on pages 16–26 Answers on pages 16–26

4 Differentiate with respect to x

(a) $(x^2 + 1)\sqrt{x}$, [4]

(b) $\dfrac{x^2 + x^9}{x^7}$. [4]

5 The point with coordinates (1, 5) lies on the curve C with equation $y = x^3 + ax^2 + bx$, where a and b are constants. The equation of the tangent to the curve C at the point with coordinates (1, 5) is $y = 2x + 3$. Find the values of a and b. [6]

6 **(a)** Find the coordinates of the points on the curve with equation

$$y = 2x^3 + x^2 - x + 2$$

at which the tangents are parallel to the line with equation $y = 3x + 2$. [7]

(b) Find the equation of the normal to the curve at the point with coordinates (1, 4). [4]

7 **(a)** Show that the equation of the tangent to the curve with equation $y = x^3 + x$ at the point where $x = 1$ is $y = 4x - 2$. [4]

This tangent meets the curve again at the point where $x = a$.

(b) Show that $a^3 - 3a + 2 = 0$. [4]

(c) Hence find the value of a. [4]

8 The curve C has equation $y = f(x)$ where

$$f(x) = x^3 + 4x - \frac{3}{x}, \; x \neq 0.$$

The points P and Q both lie on C and have coordinates (1, 2) and (−1, −2) respectively.

(a) Show that the curve C has the same gradient at P and Q. [5]

(b) Find the equation of the tangent to C at P. [4]

The tangent to C at P meets the x-axis at the point A and meets the y-axis at the point B.

(c) Find the area of triangle OAB. [4]

1.5 Integration

1 **(a)** Find the general solution of the differential equation

$$\frac{dy}{dx} = 2(\sqrt{x} - 1)^2.$$

Given that when $x = 9$, $y = 2$, [5]

(b) find y in terms of x.

Answers on pages 16–26 **Answers** on pages 16–26 **Answers** on pages 16–26

2 The curve C has equation $y = f(x)$ and passes through the point $P(2, 4)$.
Given that $f'(x) = 3x^2 - 6x - 4$,

(a) find $f(x)$. [4]

The point $Q(-1, 16)$ lies on C.
The tangent to C at Q is parallel to the tangent to C at R.

(b) Find the coordinates of R. [6]

3 Given $y = (4 - \sqrt{x})$, find:

(a) $\int y dx$, [2]

(b) $\int y^2 dx$. [4]

4 Given that $\frac{dy}{dx} = 3x^{\frac{1}{2}} - 2x^{-\frac{1}{2}}$ and that $y = 13$ when $x = 4$,

(a) find y in terms of x. [4]

(b) Hence find the value of y when $x = 9$. [2]

5 Given $y = \frac{2x^4 + x - 3}{x^3}$ find:

(a) $\int y dx$, [4]

(b) $\frac{dy}{dx}$. [2]

6 The curve C has equation $y = 2x^3 - 9x^2 - 12x + 1$.

(a) Find the coordinates of the points on C at which the gradient is 12. [7]

(b) Find $\int y dx$. [2]

7 $y = (x^{\frac{1}{2}} + x^{-\frac{1}{2}})(x^{\frac{1}{2}} - x^{-\frac{1}{2}})x$

(a) Express y in the form $ax^2 + b$, where a and b are constants to be found. [2]

Hence find:

(b) $\int y dx$, [2]

(c) $\frac{dy}{dx}$. [2]

8 Find:

(a) $\int x^{-\frac{1}{4}} dx$. [1]

(b) $\int \left(x - \frac{3}{x}\right)^2 dx$. [4]

Answers on pages 16–26 **Answers** on pages 16–26 **Answers** on pages 16–26

Answers

1.1 Algebra and functions

examiner's tips

(1) Equation must factorise as

$$\{x - (2 + \sqrt{5})\}\{x - (2 - \sqrt{5})\} = 0$$

$$x\{x - (2 - \sqrt{5})\} - (2 + \sqrt{5})\{x - (2 - \sqrt{5})\} = 0$$ ← **Multiply out the brackets.**

$$x^2 - x(2 - \sqrt{5} + 2 + \sqrt{5}) + (2 - \sqrt{5})(2 + \sqrt{5}) = 0$$

$$x^2 - 4x - 1 = 0$$ ← **Collect terms.**

So, $b = -4$ and $c = -1$ ← **Equating coefficients.**

(2) $y = 2x + 1$ ← **It is much easier to substitute for *y*.**

$x(2x + 1) + 2x + 1 = 3$

$2x^2 + 3x - 2 = 0$ ← **Multiply out and collect terms.**

$(x + 2)(2x - 1) = 0$ ← **Factorise or use formula.**

$x = -2, y = -3$

or $x = \frac{1}{2}, y = 2$.

(3) (a) $y = 2^x \Rightarrow 4y^2 = 2^2.(2^x)^2 = 2^2.(2^{2x}) = 2^{2x+2}$

(b) So, $2^{2x+2} + 7.2^x - 2 = 0$ ← **Now put $y = 2^x$.**

$\Rightarrow 4y^2 + 7y - 2 = 0$ ← **Using the result from (a).**

$\Rightarrow (4y - 1)(y + 2) = 0$ ← **Always factorise if possible.**

$\Rightarrow y = \frac{1}{4} \Rightarrow 2^x = \frac{1}{4} \Rightarrow x = -2$

or $y = -2$ no solution for x ← **Note that $2^x = -2$ has no solution as $2^x > 0$ for all x.**

(4) $$\frac{2}{(\sqrt{2} - 1)} + \frac{3}{(\sqrt{2} + 1)} = \frac{2(\sqrt{2} + 1) + 3(\sqrt{2} - 1)}{(\sqrt{2} - 1)(\sqrt{2} + 1)}$$ ← **Using the common denominator.**

$$= \frac{2\sqrt{2} + 2 + 3\sqrt{2} - 3}{(2 - 1)}$$ ← **Multiplying out.**

$$= 5\sqrt{2} - 1$$ ← **Collecting the terms.**

(5) (a) $2x^2 + 8x - 1 = 2(x^2 + 4x) - 1$

$= 2(x^2 + 4x + 4) - 8 - 1$ ← **Completing the square as required.**

$= 2(x + 2)^2 - 9$

So $a = 2$, $b = 2$ and $c = -9$

(b) $2x^2 + 8x - 1 = 0$

$\Rightarrow 2(x + 2)^2 - 9 = 0$ ← **From part (a).**

$\Rightarrow 2(x + 2)^2 = 9$

$\Rightarrow (x + 2)^2 = \frac{9}{2}$ ← **Now square root both sides.**

$\Rightarrow x = \pm\frac{3}{\sqrt{2}} - 2 = -2 \pm \frac{3}{2}\sqrt{2}$ ← **Two square roots!**

(c) Since $y = 2(x + 2)^2 - 9$, minimum point on graph is $(-2, -9)$. Also graph crosses x-axis (twice) at $x = -2 \pm \frac{3}{2}\sqrt{2}$ ← **From part (b).**

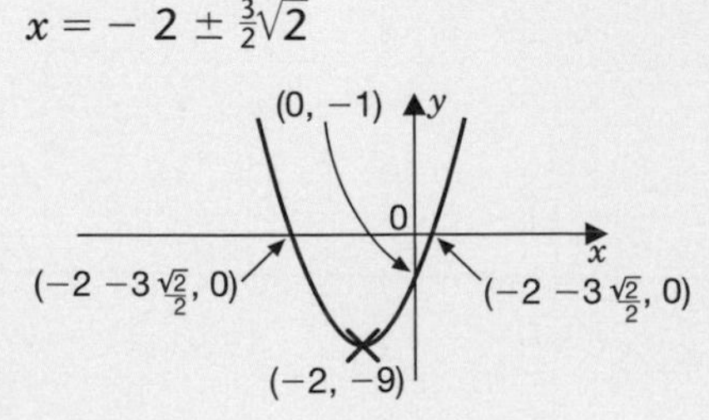

(d) (i) $2x^2 + 8x - 1 = -9 \Rightarrow x = -2$ ← **From part (c).**

(ii) $2x^2 < 1 - 8x \Rightarrow 2x^2 + 8x - 1 < 0$

$\Rightarrow -2 - \frac{3}{2}\sqrt{2} < x < -2 + \frac{3}{2}\sqrt{2}$

Core 1

(6) (a) $12x + 4y = 160$ ← Total perimeter is 160 m.

$3x + y = 40$ ← Divide by 4.

(b) $3x^2 + y^2 = 508$ ← Total area is $508\ \text{m}^2$.

(c)

$$y = 40 - 3x$$ ← Easier to substitute for y here.

$$3x^2 + (40 - 3x)^2 = 508$$

$$3x^2 + 1600 - 240x + 9x^2 = 508$$ ← Multiplying out.

$$12x^2 - 240x + 1092 = 0$$ ← Collecting the terms.

$$x^2 - 20x + 91 = 0$$ ← Dividing through by 12.

$$(x - 7)(x - 13) = 0$$ ← Factorising.

$$x = 7 \quad \text{or} \quad x = 13$$ ← Two solutions for x.

$$y = 19 \quad \text{or} \quad y = 1$$ ← Corresponding values of y.

So $x = 7$ and $y = 19$ ← This is the only solution as $y > x$ (given in the question).

(7) (a) $(x^{\frac{1}{2}} + x^{-\frac{1}{2}})(x^{\frac{1}{2}} - x^{-\frac{1}{2}})$ ← This is a difference of two squares.

$$= x - x^{-1}$$

$$= x - \frac{1}{x}$$

$$= \frac{x^2 - 1}{x}$$ ← Using a common denominator.

(b) $\frac{8}{3} = \frac{3^2 - 1}{3}$

$$= (3^{\frac{1}{2}} + 3^{-\frac{1}{2}})(3^{\frac{1}{2}} - 3^{-\frac{1}{2}})$$ ← Using part (a).

(8) (a) $6x - 5 - x^2 = -(x^2 - 6x) - 5$

$$= -(x^2 - 6x + 9) - 5 + 9$$

$$= -(x - 3)^2 + 4$$ ← Completing the square as required.

So $p = -1$, $q = -3$ and $r = 4$.

(b) Since $(x - 3)^2 \geqslant 0$, maximum value is 4 when $x = 3$.

(c) Put $y = 0 : 0 = 6x - 5 - x^2$ ← Put $y = 0$ to find where the graph crosses the x-axis.

$$\Rightarrow x^2 - 6x + 5 = 0$$

$$\Rightarrow (x - 5)(x - 1) = 0$$

$$\Rightarrow x = 5 \text{ or } 1$$

Also, when $x = 0$, $y = -5$

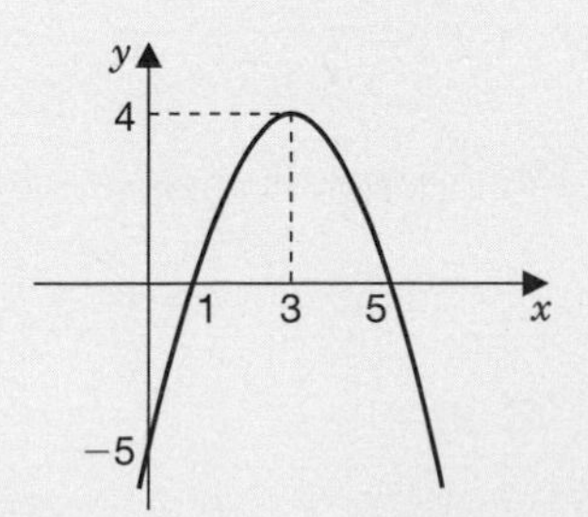

← Maximum point is (3, 4) from part (b).

(9) (a) For two real distinct solutions, $b^2 > 4ac$ ← You need to learn this.

$$\Rightarrow (k + 1)^2 > 4$$

(b) $\Rightarrow k + 1 > 2 \quad \text{or} \quad k + 1 < -2$ ← Care must be taken when square rooting an inequality.

$$\Rightarrow k > 1 \quad \text{or} \quad k < -3$$

Core 1

(10) (a) $2x^2 - 4x + 1 = 0$

$$x = \frac{4 \pm \sqrt{16 - 8}}{4}$$
$$= 1 \pm \frac{2\sqrt{2}}{4}$$
$$= 1 \pm \tfrac{1}{2}\sqrt{2}$$

Using the quadratic formula. N.B. $\sqrt{8} = \sqrt{4 \times 2} = 2\sqrt{2}$

(b) (i) Sum of roots $= 1 + \tfrac{1}{2}\sqrt{2} + 1 - \tfrac{1}{2}\sqrt{2}$
$= 2$

Simplifying.

(ii) Product of roots $= (1 + \tfrac{1}{2}\sqrt{2})(1 - \tfrac{1}{2}\sqrt{2})$
$= 1 - \tfrac{1}{4} \times 2$
$= \tfrac{1}{2}$

Multiplying out.

(11) (a)

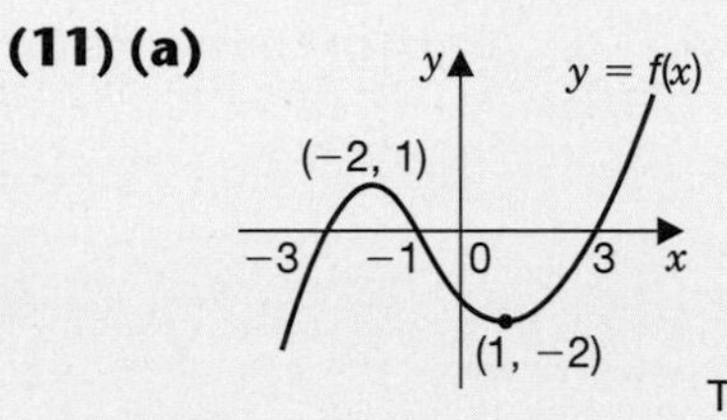

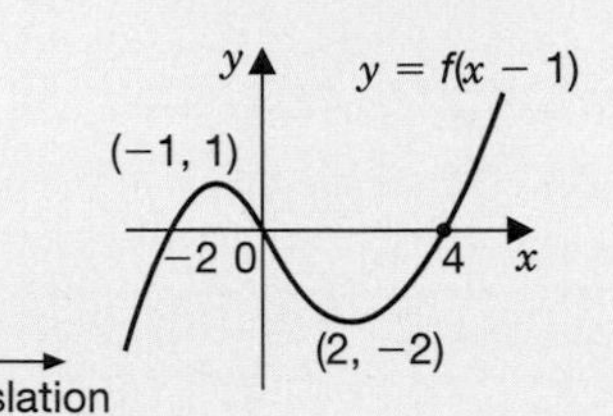

Translation

Note that changing x to $x - 1$ translates the graph by 1 unit in the +ve x-direction.

(b)

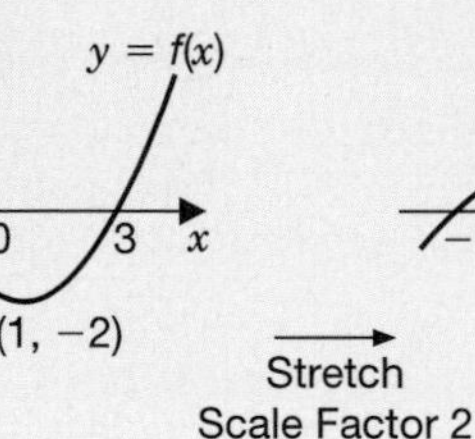

Stretch
Scale Factor 2

Here changing x to $\tfrac{1}{2}x$ stretches the graph by a scale factor of 2 in the x-direction.

1.2 Coordinate geometry

(1) (a) Mid-point of AB is $\left(\frac{-4 + -2}{2}, \frac{2 + 4}{2}\right)$ i.e. $(-3, 3)$;

therefore the gradient of $l = \frac{3}{-3} = -1$.

Hence, an equation of l is $y = -x$.

Notice that 'an' is used since it isn't the only one, e.g. $2y = -2x$.

(b) Mid-point of OA is $(-2, 1)$; hence the line m has equation $x = -2$.

(c) Solving $y = -x$ and $x = -2$ simultaneously gives the point $(-2, 2)$.

(2) (a) $AD = \sqrt{(0 - 6)^2 + (18 - 15)^2} = \sqrt{45} = 6.71$ (3 s.f.)

(b) Gradient of $AD = \frac{(18 - 15)}{(0 - 6)} = -\tfrac{1}{2}$;

equation of l is $y - 18 = -\tfrac{1}{2}(x - 0)$ i.e. $y = -\tfrac{1}{2}x + 18$.

(c) Since C lies on the curve, $16 = 16.4 - 16k \rightarrow k = 3$.

(d) B lies on both the line and the curve hence, $-\tfrac{1}{2}x + 18 = 16x - 3x^2$

$$6x^2 - 33x + 36 = 0$$
$$2x^2 - 11x + 12 = 0$$
$$(x - 4)(2x - 3) = 0$$
$$x = 4 \text{ or } x = \tfrac{3}{2}$$

When $x = \tfrac{3}{2}$, $y = 17\tfrac{1}{4}$, B is $(\tfrac{3}{2}, 17\tfrac{1}{4})$

Note that $(x - 4)$ must be a factor as C has x-coordinate of 4.

(3) $2x + 3y - 4 = 0$

$\Rightarrow y = -\frac{2}{3}x + \frac{4}{3}$ — **Rearrange in form $y = mx + c$.**

$\Rightarrow$ gradient is $-\frac{2}{3}$ — **m is the gradient.**

Equation is $y - 3 = -\frac{2}{3}(x - -1)$ — **Using $y - y_1 = m(x - x_1)$.**

$\Rightarrow \quad 3y - 9 = -2x - 2$ — **Multiplying through by 3.**

$\Rightarrow \quad 3y + 2x - 7 = 0$

i.e. $2x + 3y - 7 = 0$ — **As required – note that $-2x - 3y + 7 = 0$ is equally good.**

(4) (a) M is $\left(\dfrac{-2 + 4}{2}, \dfrac{5 + 3}{2}\right)$ — **Mid-point is given by the mean of the two points.**

i.e. (1, 4)

(b) Gradient of $AB = \dfrac{5 - 3}{-2 - 4}$ — **Using $\dfrac{y_1 - y_2}{x_1 - x_2}$ for the gradient.**

$= -\frac{1}{3}$

Gradient of perpendicular is 3. — **To find the perpendicular gradient 'flip the number and change its sign'.**

So, equation of bisector is

$y - 4 = 3(x - 1)$ — **Using answer to part (a).**

$\Rightarrow \quad y = 3x + 1$

(c) When $x = 2$, $y = 7$

(d) Perpendicular distance is PM

i.e. $\sqrt{(2 - 1)^2 + (7 - 4)^2}$

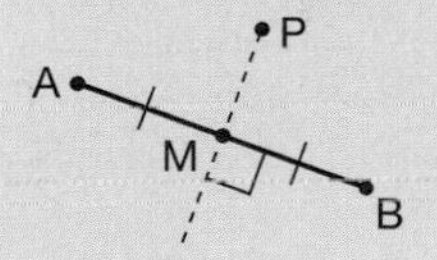

Distance is PM.

$= \sqrt{1 + 9}$

$= \sqrt{10}$

(5) (a) $3x + 2y + 3 = 0$

$\Rightarrow y = -\frac{3}{2}x - \frac{3}{2}$ — **Put equation in $y = mx + c$ form.**

$\Rightarrow$ gradient is $-\frac{3}{2}$ — **Gradient is m.**

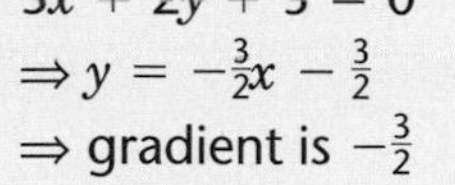

(b) $\tan \theta = -\frac{3}{2}$

$\Rightarrow \theta = -56^0$ — **Angle with negative x-axis is given by $\tan \theta = m$.**

L makes an angle of 124° with the positive x-axis.

(6) (a) $L_1 : 2x + 4y + 1 = 0 \Rightarrow y = -\frac{1}{2}x - \frac{1}{4}$

$\therefore$ gradient of L_1 is $-\frac{1}{2}$

$L_2 : 2y = 3 - x \Rightarrow y = \frac{3}{2} - \frac{1}{2}x$

$\therefore$ gradient of L_2 is $-\frac{1}{2}$

$\therefore$ lines L_1 and L_2 are parallel.

(b) $(2 \times -2.5) + 4 + 1 = 0$ — **Hence A lies on L_1.**

$2 \times 2.4 = 3 - -1.8$ — **Hence B lies on L_2.**

(c) Gradient of $AB = \dfrac{2.4 - 1}{-1.8 - -2.5}$ — **Using $\dfrac{y_2 - y_1}{x_2 - x_1}$.**

$= \dfrac{1.4}{0.7} = 2$

Gradient of $L_1 = -\frac{1}{2}$

As, $2 \times -\frac{1}{2} = -1$, lines are perpendicular.

Core 1

(d) Perpendicular distance $= AB$

← **Since L_1 and L_2 are parallel, AB is also perpendicular to L_2.**

$= \sqrt{(-2.5 - -1.8)^2 + (1 - 2.4)^2}$ ← **Using $\sqrt{(x_1 - x_2)^2 + (y_1 - y_2)^2}$.**

$= \sqrt{0.7^2 + 1.4^2}$

$= \sqrt{0.7^2(1^2 + 2^2)}$ ← **Taking out the common factor.**

$= 0.7\sqrt{5}$

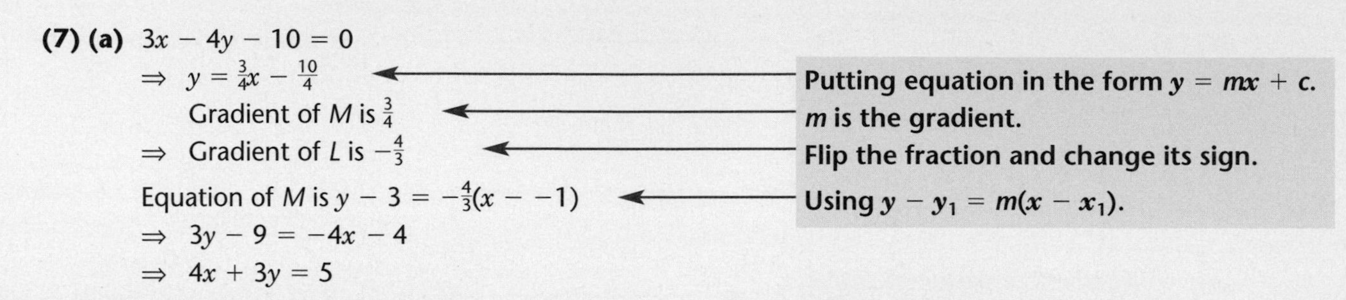

(7) (a) $3x - 4y - 10 = 0$

$\Rightarrow\ y = \frac{3}{4}x - \frac{10}{4}$ ← **Putting equation in the form $y = mx + c$.**

Gradient of M is $\frac{3}{4}$ ← **m is the gradient.**

$\Rightarrow$ Gradient of L is $-\frac{4}{3}$ ← **Flip the fraction and change its sign.**

Equation of M is $y - 3 = -\frac{4}{3}(x - -1)$ ← **Using $y - y_1 = m(x - x_1)$.**

$\Rightarrow\ 3y - 9 = -4x - 4$

$\Rightarrow\ 4x + 3y = 5$

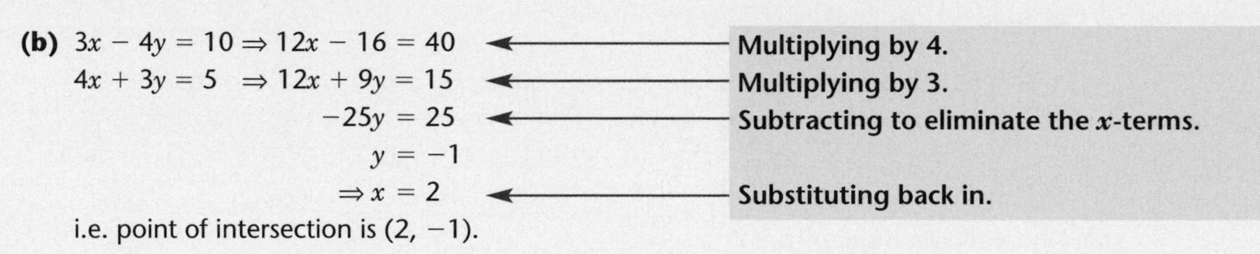

(b) $3x - 4y = 10 \Rightarrow 12x - 16 = 40$ ← **Multiplying by 4.**

$4x + 3y = 5 \Rightarrow 12x + 9y = 15$ ← **Multiplying by 3.**

$-25y = 25$ ← **Subtracting to eliminate the x-terms.**

$y = -1$

$\Rightarrow x = 2$ ← **Substituting back in.**

i.e. point of intersection is $(2, -1)$.

1.3 Sequences and series

(1) (a) $a + 2d = -20$ ← **nth term $= a + (n - 1)d$ – you need to learn this!**

$a + 10d = 20$

subtracting, $8d = 40 \rightarrow d = 5 \rightarrow a = -30$.

(b) $\frac{k}{2}\{2a + (k - 1)d\} = 0$ ← **You need to know this.**

$\frac{k}{2}\{-60 + 5(k - 1)\} = 0$ ← **Substituting for a and d.**

$\frac{k}{2}(5k - 65) = 0$ ← **Do not multiply out!**

$5k - 65 = 0$ since $k \neq 0$

$k = 13$

(2) (a) The repayments form an AP with $a = 20$ and $d = 2$

nth term $= a + (n - 1)d = 114$

$20 + (n - 1)2 = 114$ ← **Now make n the subject.**

$2(n - 1) = 94$

$(n - 1) = 47$

$n = 48$

(b) The price of the car is the sum of the 48 repayments as the loan is interest-free

i.e. $P = S_{48} = \frac{(20 + 114)}{2} \times 48$ ← **Here it is easier to use $Sn = \frac{(a + l)n}{2}$.**

$= 67 \times 48 = £3216$

Core 1

(3) (a) $u_1 = 1, u_2 = \frac{1}{3}, u_3 = -\frac{1}{9}$

(b) $\sum_{n=1}^{15} 3(\frac{2}{3})^n = 2 + \frac{4}{3} + \frac{8}{9} + \ldots$ ← **This is a GP with common ratio $\frac{2}{3}$ to 15 terms.**

$= 2\,(1 - (\frac{2}{3})^{15})/1 - \frac{2}{3})$ ← **You need to know $Sn = a(1 - r^n)/(1 - r)$.**

$= 6\,(1 - (\frac{2}{3})^{15}) = 5.986$ (4 s.f.)

$\sum_{n=1}^{15} u_n = \sum_{n=1}^{15} \left\{3(\frac{2}{3})^n - 1\right\} = \sum_{n=1}^{15} 3(\frac{2}{3})^n - 15 = 5.986 - 15 = -9.014$ (4 s.f.).

(c) $LHS = 3\{3(\frac{2}{3})^{n+1} - 1\} = 3^2(2^{n+1}\frac{1}{3}^{n+1}) - 3 = 3^{1-n}\,2^{n+1} - 3$

$RHS = 2\{3(\frac{2}{3})^n - 1\} - 1 = 3^{1-n}\,2^{n+1} - 2 - 1 = 3^{1-n}\,2^{n+1} - 3 = LHS.$

(4) (a) (i) $\sum_{r=1}^{50} (3r - 7)$

$= -4 + -1 + 2 + \ldots + 143$ ← **Write out the first few terms to decide what type of series it is. Using $S = \frac{(a + l)}{2}n$.**

$= \frac{(-4 + 143)}{2} \times 50$

$= 3475$

(ii) $\sum_{r=1}^{30} (3r - 7) = -4 + \ldots + 83$

$= \frac{(-4 + 83)}{2} \times 30$ ← **As above.**

$= 1185$

(b) Hence, $\sum_{r=31}^{50} (3r - 7)$ ← **$\sum_{r=31}^{50} (3r - 7) = \sum_{r=1}^{50} (3r - 7) - \sum_{r=1}^{30} (3r - 7)$.**

$= 3475 - 1185$

$= 2290$

(5) (a) $S_3 = 3a + 9b = 18$ ← **Putting $n = 3$ in the given formula.**

$S_5 = 5a + 25b = 50$ **Putting $n = 5$ in the given formula.**

$\Rightarrow 15a + 45b = 90$

$\Rightarrow 15a + 75b = 150$ **Subtracting.**

$30b = 60$

$b = 2$

$\Rightarrow a = 0$

(b) nth term $= S_n - S_{n-1}$ ← **Using $a = 0$, $b = 2$.**

$= 2n^2 - 2(n - 1)^2$ ← **Multiplying out.**

$= 2n^2 - 2(n^2 - 2n + 1)$

$= 2n^2 - 2n^2 + 4n - 2$

$= 4n - 2$

(c) $u_1 = 4 - 2 = 2$ ← **Putting $n = 1$.**

$u_2 = 8 - 2 = 6$ ← **Putting $n = 2$.**

$\therefore$ Common difference is 4

(6) (a) First term, $a = \frac{1}{2} \times 1 \times (3 + 5)$
$= 4$ ← **Putting $n = 1$.**

(b) $S_2 = \frac{1}{2} \times 2 \times (6 + 5)$ ← **Putting $n = 2$.**
$= 11$

so, second term $= 11 - 4 = 7$ ← **$u_2 = S_2 - S_1$.**

common difference $= 7 - 4 = 3$

(c) $S_{20} = \frac{20}{2}((3 \times 20) + 5) = 650$ ← **Putting $n = 20$.**

(7) (a) Let n be number of terms.

$$\left(\frac{-2 + 73}{2}\right)n = 923$$
$$n = 26$$
← **Using $S_n = \frac{(a + b)}{2}n$.**

(b) $-2 + 25d = 73$
$d = 3$ ← **Using nth term $= a + (n - 1)d$.**

(8) (a) $S_n = \frac{n}{2}\{4 + (n - 1)3\}$ ← **Using $S_n = \frac{n}{2}\{2a + (n - 1)d\}$.**
$= \frac{n}{2}(3n + 1)$

(b) $\frac{n}{2}(3n + 1) = 610$ ← **From part (a).**

$3n^2 + n - 1220 = 0$ ← **Collecting terms after multiplying by 2.**

$(3n + 61)(n - 20) = 0$ ← **Factorising (or use formula).**

$\Rightarrow \quad n = -\frac{61}{3}$ or $n = 20$

so, 20 terms ← **Neglecting the negative value.**

(9) (a) £100 + (35 × 20) = £800 ← **nth term $= a + (n - 1)d$.**

(b) Total paid $= \frac{36}{2}\{200 + (35 \times 20)\}$ ← **Using $\frac{n}{2}\{2a + (n - 1)d\}$.**
$= 18 \times 900$
$=$ £16 200

∴ Interest $= 16\,200 - 10\,000$
$=$ £6200

1.4 Differentiation

(1) (a) $\frac{dy}{dx} = 2 - 2x$

When $x = 0$, gradient of tangent is 2

$\rightarrow$ gradient of normal is $-\frac{1}{2}$ ← **Change the sign and invert to obtain the gradient of normal.**

equation of normal at O is $y = -\frac{1}{2}x$.

(b) At intersections of normal and curve,

$-\frac{1}{2}x = 2x - x^2$ i.e. $0 = 5x - 2x^2$ ← **Solving the normal and curve equations simultaneously.**
$0 = x(5 - 2x)$
$x = 0$ or $\frac{5}{2}$

When $x = \frac{5}{2}$, $y = 5 - \left(\frac{5}{2}\right)^2 = -\frac{5}{4}$

The point is $\left(\frac{5}{2}, -\frac{5}{4}\right)$.

Core 1

(2) (a) $y = (x^{\frac{1}{2}} + x^{\frac{3}{2}})^2$ ← We need to multiply out before differentiation.

$\Rightarrow \quad y = x + 2x^2 + x^3$

$\Rightarrow \quad \dfrac{dy}{dx} = 1 + 4x + 3x^2$

(b) $\dfrac{d^2y}{dx^2} = 4 + 6x = 0$ ← Equating the second derivative to zero.

$\Rightarrow \quad x = -\frac{2}{3}$

(3) $-1 = 4a + 2b + c \quad (1)$ ← The point $(2, -1)$ lies on the curve.

$\dfrac{dy}{dx} = 2ax + b$

so, $\quad 3 = 4a + b \quad (2)$ ← $\dfrac{dy}{dx} = 3$ when $x = 2$.

$\dfrac{d^2y}{dx^2} = 2a$

so, $\quad -2 = 2a \Rightarrow a = -1$ ← We now substitute back into equations (1) and (2).

$3 = -4 + b \Rightarrow b = 7$

$-1 = -4 + 14 + c \Rightarrow c = -11$

(4) (a) $(x^2 + 1)x^{\frac{1}{2}} = x^{\frac{5}{2}} + x^{\frac{1}{2}}$ ← Multiply out first.

$\dfrac{d}{dx}(x^{\frac{5}{2}} + x^{\frac{1}{2}}) = \dfrac{5}{2}x^{\frac{3}{2}} + \dfrac{1}{2}x^{-\frac{1}{2}}$ ← Differentiating.

(b) $\dfrac{{}^{2}x + x^9}{x\ {}^{7}} = x^{-5} + x^2$ ← Divide out first.

$\dfrac{d}{dx}(x^{-5} + x^2) = -5x^{-6} + 2x$ ← Differentiating.

(5) $5 = 1 + a + b$ ← $(1, 5)$ lies on the curve.

$\Rightarrow \quad 4 = a + b \quad (1)$

$\dfrac{dy}{dx} = 3x^2 + 2ax + b$

When $x = 1$, $\dfrac{dy}{dx} = 3 + 2a + b$

so, $\quad 3 + 2a + b = 2$ ← The gradient of the tangent is 2.

$\Rightarrow \quad 2a + b = -1 \quad (2)$ ← We now solve (1) and (2) simultaneously.

$(2) - (1)$: $a = -5 \Rightarrow b = 9$

(6) (a) $y = 2x^3 + x^2 - x + 2$

$\dfrac{dy}{dx} = 6x^2 + 2x - 1$ ← This gives the gradient of the curve.

The line $y = 3x + 2$ has gradient 3 ← Using $y = mx + c$.

Hence, $\quad 6x^2 + 2x - 1 = 3$ ← Parallel lines have equal gradients.

$\Rightarrow \quad 3x^2 + x - 2 = 0$ ← Dividing by 2.

$\Rightarrow \quad (3x - 2)(x + 1) = 0$ ← Factorise (or use the formal).

$\Rightarrow \quad x = \frac{2}{3}$ and $y = \frac{64}{27}$ ← Now substitute these values back into the equation of the curve.

$\Rightarrow \quad x = -1$ or $y = 2$

Hence, the points are $(\frac{2}{3}, \frac{64}{27})$ and $(-1, 2)$

(b) When $x = 1$, $\dfrac{dx}{dy} = 6 + 2 - 1$

$= 7$

gradient of normal is $-\frac{1}{7}$ ← Flip the fraction and change the sign.

Equation of normal is $y - 4 = -\frac{1}{7}(x - 1)$ ← Using $y - y_1 = m(x - x_1)$.

$\Rightarrow \quad 7y - 28 = -x + 1$

$\Rightarrow \quad 7y + x - 29 = 0$

Core 1

(7) (a) $y = x^3 + x$; when $x = 1, y = 2$ — We need the y-coordinate of the point.

$\frac{dy}{dx} = 3x^2 + 1$

When $x = 1, \frac{dy}{dx} = 4$ — This gives the gradient of the tangent.

Equation of tangent is

$y - 2 = 4(x - 1)$ — Using $y - y_1 = m(x = x_1)$.

$y = 4x - 2$

(b) The tangent meets the curve again when

$x^3 + x = 4x - 2$ — Equating the y-coordinates.

$\Rightarrow x^3 - 3x + 2 = 0$ i.e. $a^3 - 3a + 2 = 0$

(c) Two of the roots of this cubic equation are 1 (where the tangent was drawn) — The tangent touches the curve at $x = 1$.

i.e. $(a - 1)^2(a + 2) = 0$

Hence, the third root is $a = -2$.

(8) (a) $f(x) = x^3 + 4x - 3x^{-1}$ — Put all the terms in the form x^n.

$f'(x) = 3x^2 + 4 + 3x^{-2}$ — Differentiate.

Hence, when $x = \pm 1$,

$f'(x) = 3 + 4 + 3$

$= 10$

(b) Equation of tangent is

$y - 2 = 10(x - 1)$ — Using $y - y_1 = m(x - x_1)$.

$y = 10x - 8$

(c) When $y = 0$, $10x - 8 = 0$ — To find A.

$\Rightarrow x = \frac{4}{5}$ i.e. A is $(\frac{4}{5}, 0)$

When $x = 0, y = -8$ — To find B.

i.e. B is $(0, -8)$

Area of $\triangle OAB = \frac{1}{2} \cdot \frac{4}{5} \cdot 8$ — $\frac{1}{2}$base × height.

$= \frac{16}{5}$ square units

1.5 Integration

(1) (a) $\frac{dy}{dx} = 2(x - \sqrt{x} + 1)$ — Multiply out the brackets.

$= 2(x - 2x^{0.5} + 1)$ — Express x terms as powers before integrating.

$y = \int(2x - 4x^{0.5} + 2)dx$

$= x^2 - \frac{8}{3}x^{1.5} + 2x + c$ — This is the general solution which must include c.

(b) $2 = 81 - 72 + 18 + c$ — This corresponds to the curve passing through (9, 2).

$\Rightarrow c = -25$

so $y = x^2 - \frac{8}{3}x^{1.5} + 2x - 25$

Core 1

(2) (a) $f'(x) = 3x^2 - 6x - 4$

$f(x) = \int(3x^2 - 6x - 4)dx$ ← Integrate to find f.

$= x^3 - 3x^2 - 4x + c$ ← Don't forget the constant.

$f(2) = 4$, since P lies on c

$4 = 2^3 - 3(2)^2 - 4(2) + c$ ← Use the coordinates of P to find the constant.

$16 = c$

so, $f(x) = x^3 - 3x^2 - 4x + 16$

(b) $f'(-1) = 3 + 6 - 4 = 5$ ← This is the gradient of C at Q.

so, $f'(x) = 5$ at R ← Since tangents are parallel.

$\Rightarrow \quad 3x^2 - 6x - 4 = 5$

$\Rightarrow \quad 3x^2 - 6x - 9 = 0$

$\Rightarrow \quad x^2 - 2x - 3 = 0$ ← Collecting terms and dividing by 3.

$\Rightarrow \quad (x - 3)(x + 1) = 0$

$\Rightarrow \quad x = 3 \quad \text{or} \quad x = -1$ ← $x = -1$ is Q!

at R, $y = f(3)$ ← Find the y-coordinate of R.

$= 27 - 27 - 12 + 16$

$= 4$

i.e. R is $(3, 4)$

(3) (a) $\int(4 - x^{\frac{1}{2}})dx$ ← Write $\sqrt{x}$ as $x^{\frac{1}{2}}$ first.

$= 4x - \frac{2}{3}x^{\frac{3}{2}} + c$ ← Don't forget the constant.

(b) $y^2 = (4 - x^{\frac{1}{2}})^2$ ← Multiply out first.

$= 16 - 8x^{\frac{1}{2}} + x$

$\int(16 - 8x^{\frac{1}{2}} + x)dx$

$= 16x - \frac{16}{3}x^{\frac{3}{2}} + \frac{x^2}{2} + c$

(4) (a) $\frac{dy}{dx} = 3x^{\frac{1}{2}} - 2x^{-\frac{1}{2}}$

$y = \int(3x^{\frac{1}{2}} - 2x^{-\frac{1}{2}})dx$ ← We need to integrate.

$= 2x^{\frac{3}{2}} - 4x^{\frac{1}{2}} + c$

$13 = 2.4^{\frac{3}{2}} - 4.4^{\frac{1}{2}} + c$ ← Use the point to find c.

$13 = 16 - 8 + c$

$5 = c$

$y = 2x^{\frac{3}{2}} - 4x^{\frac{1}{2}} + 5$

(b) When $x = 9$, $y = 2.9^{\frac{3}{2}} - 4.9^{\frac{1}{2}} + 5$ ← No calculators are allowed on C1!

$= 54 - 12 + 5$

$= 47$

(5) (a) $y = \dfrac{2x^4 + x - 3}{x^3}$

$= 2x + x^{-2} - 3x^{-3}$ ← **Divide out first.**

$\int y\,dx = \int (2x + x^{-2} - 3x^{-3})dx$ ← **Integrating.**

$= x^2 - x^{-1} + \dfrac{3}{2}x^{-2} + c$

(b) $y = 2x + x^{-2} - 3x^{-3}$

$\dfrac{dy}{dx} = 2 - 2x^{-3} + 9x^{-4}$

(6) (a) $\dfrac{dy}{dx} = 6x^2 - 18x - 12$ ← **Differentiate to find the gradient.**

$6x^2 - 18x - 12 = 12$

$x^2 - 3x - 4 = 0$ ← **Divide through by 6.**

$(x - 4)(x + 1) = 0$ ← **Factorising.**

$\Rightarrow\ x = 4$ and $y = -63$

$\Rightarrow\ x = -1$ and $y = 2$ ← **We need the y-coordinate also.**

(b) $\int (2x^3 - 9x^2 - 12x + 1)dx$

$= \frac{1}{2}x^4 - 3x^3 - 6x^2 + x + c$ ← **Don't forget the constant.**

(7) (a) $y = (x^{\frac{1}{2}} + x^{-\frac{1}{2}})(x^{\frac{1}{2}} - x^{-\frac{1}{2}})x$ ← **The brackets give a difference of two squares.**

$= (x - x^{-1})x = x^2 - 1$ ← **Multiplying out.**

i.e. $a = 1$; $b = -1$

(b) $\int y\,dx = \int (x^2 - 1)dx$ ← **Now integrate.**

$= \frac{1}{3}x^3 - x + c$

(c) $\dfrac{dy}{dx} = 2x$

(8) (a) $\int x^{-\frac{1}{4}}\,dx = \frac{4}{3}x^{\frac{3}{4}} + c.$ ← **It is best to use top-heavy fractions.**

(b) $\int \left(x - \dfrac{3}{x}\right)^2 dx$

$= \int \left(x^2 - 6 + \dfrac{9}{x^2}\right)dx$ ← **Multiply out brackets first.**

$= \int (x^2 - 6 + 9x^{-2})dx$ ← **Express each term as a power of x before integrating.**

$= \frac{1}{3}x^3 - 6x - 9x^{-1} + c.$ ← **Always include an arbitrary constant.**

Questions with model answers

C grade candidate – mark scored 8/15

(1) (a) Find the coefficient of x^2 in the expansion of $(3 - 2x)^5$. [4]

Term is $\binom{5}{2} -2x^2 \; 3^3$

Hence coeff. $= 10 \times -2 \times 27 = -540$

(b) Find the constant term in the expansion of $\left(x^2 - \frac{1}{x}\right)^6$. [4]

Term is $\binom{6}{2} (x^2)^2 \left(-\frac{1}{x}\right)^4$

i.e. $\frac{6 \times 5}{1 \times 2} = 15.$

Given that the coefficients of y, x^2 and x^3 in the expansion of $(1 + x)^n$, where $n \geqslant 3$, are in arithmetic progression,

(c) find the value of n. [7]

coefficients are: $\binom{n}{1}, \binom{n}{2}, \binom{n}{3}$

so, $\binom{n}{3} - \binom{n}{2} = \binom{n}{2} - \binom{n}{1}$

N.B.

$$\frac{n(n-1)(n-2)}{6} - \frac{n(n-1)}{2} = \frac{n(n-1)}{2} - n$$

$(n-1)(n-2) - 3(n-1) = 3(n-1) - 6$

$n^2 - 9n + 14 = 0$

$(n-7)(n-2) = 0$

$n = 7$ (as $n = 2$ is not possible since $n \geqslant 3$)

Examiner's Commentary

? For help see Revise AS Study Guide section 2

The candidate has forgotten to put brackets around $-2x$; forgetting to square the –2 is the most common error at this stage.

The binomial coefficient can be evaluated using the nCr button on a calculator, $\frac{5!}{3!2!}$ or more simply $\frac{5 \times 4}{1 \times 2}$

2/4 scored.

This ensures that the x's cancel – the brackets are remembered here!

4/4 scored.

Correct but the candidate has always relied on the calculator to evaluate binomial coefficients, cannot cope with algebraic forms and stops,

2/7 scored.

Dividing by $n(\neq 0)$ and multiplying by 6 to clear fractions.

Multiplying out and collecting all terms on one side.

It is important that a reason is given for ignoring an answer.

GRADE BOOSTER

When carrying out binomial expansions remember to use brackets (see **(a)** above) and always write out everything in full first, before simplifying.

A grade candidate – mark scored 10/12

(1) The line L with equation $y = 3x - 4$ cuts the curve C with equation $y = x^2 - 4x + 6$ at the points P and Q.

(a) Find the coordinates of the points P and Q. [5]

$x^2 - 4x + 6 = 3x - 4$ ← **Equating the *y*-coordinates.**

$x^2 - 7x + 10 = 0$

$(x - 5)(x - 2) = 0$ ← **Collecting terms and factorising.**

$x = 5$ or 2

$y = 11$ or 2

P in (2, 2) and Q in (5, 11) ← **All correct – the candidate scores 5/5.**

(b) Find the area enclosed between the curve C and the Line L. [7]

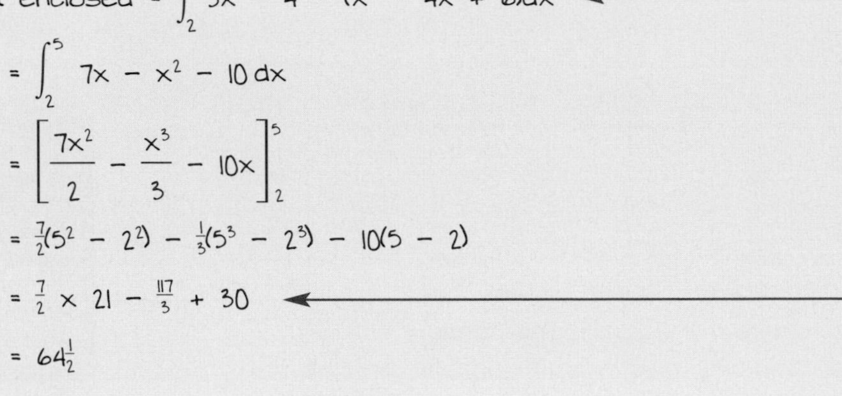

Area enclosed $= \int_2^5 3x - 4 - (x^2 - 4x + 6)dx$ ← **Correct the candidate is integrating the difference between the 2 graphs.**

$= \int_2^5 7x - x^2 - 10\,dx$

$= \left[\frac{7x^2}{2} - \frac{x^3}{3} - 10x\right]_2^5$

$= \frac{7}{2}(5^2 - 2^2) - \frac{1}{3}(5^3 - 2^3) - 10(5 - 2)$

$= \frac{7}{2} \times 21 - \frac{117}{3} + 30$ ← **Sign error here – should be – 30 not +30 5/7 scored.**

$= 64\frac{1}{2}$

Examiner's Commentary

Core 2

Exam practice questions

2.1 Algebra and functions

1 **(a)** Expand $(a + b)^4$. [4]

(b) Hence, or otherwise, simplify $x^4 + 4x^3(1 - x) + 6x^2(1 - x)^2 + 4x(1 - x)^3 + (1 - x)^4$. [4]

2 The equation of a curve is $y = ax^n$, where a and n are constants.
The points (2, 9) and (4, 15) both lie on the curve. Find the values of a and n. [9]

3 Solve

(a) $\ln(a + 10) = 2\ln(a - 2)$, [7]

(b) $2^{2x} - 6.2^x + 8 = 0$. [6]

4 The polynomial $f(x) = 2x^3 + px^2 - x + q$ is exactly divisible by $(x - 1)$ and $(x + 2)$.

(a) Find the values of p and q. [5]

(b) Hence factorise $f(x)$. [2]

5 The function f is given by

$$f(x) = px^3 + 11x^2 + 2px - 5.$$

When $f(x)$ is divided by $(x + 2)$, the remainder is 15.

(a) Show that $p = 2$. [3]

(b) Factorise $f(x)$ completely. [3]

(c) Hence find the solutions of the equation

$$f(x) = (x + 5)(x + 1).$$ [4]

6 Given that the polynomial $P(x)$ is divisible by $(x - a)^2$, show that $P'(x)$ is divisible by $(x - a)$. [4]
The polynomial $x^4 + x^3 - 12x^2 + px + q$ is divisible by $(x + 2)^2$.
Find the values of p and q. [7]

7 $P(x) = ax^3 - 3x^2 + bx + 6$, where a and b are constants.
When $P(x)$ is divided by $(x - 1)$ it leaves a remainder of -6. When divided by $(x + 2)$ it leaves a remainder of zero.

(a) Find the values of a and b. [7]

(b) Solve the equation $P(x) = 0$. [5]

Answers on pages 39–57 **Answers** on pages 39–57 **Answers** on pages 39–57

8 **(a)** Given that $(x + 4)$ is a factor of $3x^3 + x^2 + px + 24$, where p is a constant, find the value of p. [4]

(b) When divided by $(x + 2)$, the polynomial $5x^3 - 3x^2 + ax + 7$ leaves a remainder of r.
When divided by $(x + 2)$, the polynomial $4x^3 + ax^2 + 7x - 4$ leaves a remainder of $2r$.
Find the value of the constant a. [6]

9 $f(x) = ax^3 + bx^2 + cx + d$, where a, b, c and d are all constants.
The curve with equation $y = f(x)$ has gradient 4 at the point with coordinates $(0, -5)$.

(a) Find the values of c and d. [6]

(b) When $f(x)$ is divided by $(x - 1)$ it leaves a remainder of 12.
When divided by $(x + 2)$ it leaves a remainder of 15.
Find the values of a and b. [7]

10 $g(x) = 2x^3 - x^2 - 23x - 20$.

(a) Show that $(x + 1)$ is a factor of $g(x)$. [2]

(b) Factorise $g(x)$ completely. [4]

(c) Solve the equation $g(x) = 0$. [4]

2.2 Coordinate geometry

1 **(a)** Find the centre and radius of the circle with equation

$$x^2 + y^2 - 16x - 12y + 96 = 0.$$ [5]

(b) Find the greatest and least distances of the origin O from the circumference of the circle. [5]

2 The circle C passes through the points with coordinates $(0, 4)$, $(0, -4)$ and $(3, 5)$.
Find an equation for the circle C. [7]

3 **(a)** Find an equation for the circle with centre $(1, 4)$ and radius 3. [3]

(b) Determine by calculation whether the point $(2.8, 1.6)$ lies inside or outside the circle. [3]

4 **(a)** Find the centre and radius of the circle C with equation

$$x^2 + y^2 - 8x + 10y - 59 = 0$$ [5]

(b) Verify that the point P with coordinates $(10, 13)$ lies on the circle. [1]

(c) Find an equation for the tangent to C at the point P. [5]

Answers on pages 39–57 **Answers** on pages 39–57 **Answers** on pages 39–57

5 **(a)** Find the centre and radius of the circle C with equation

$$x^2 + y^2 - 8x + 24y - 9 = 0.$$ [5]

(b) Find the coordinates of the points A and B where the circle meets the line with equation $x + 1 = 0$. [4]

(c) Show that the length of AB is 24. [2]

(d) Find an equation for the tangent to C at the point $(-1, 0)$. [5]

(e) Find the x-coordinate of the point where the tangent meets the line with equation $y = -12$. [2]

6 **(a)** Find the centres and radii of the circles with equations

$$x^2 + y^2 + 10x + 20y + 25 = 0 \text{ and } x^2 + y^2 - 8x - 4y - 5 = 0.$$ [5]

(b) Deduce that the circles touch each other. [2]

7 A circle C has centre $(2, -3)$ and passes through the point $(5, -7)$. Find an equation for C. [4]

8 Given that the line $y = mx$ is a tangent to the circle with equation $x^2 + y^2 - 6x - 6y + 17 = 0$, find the possible values of m. [8]

Core 2

9 $f(x) = ax^3 + bx^2 + cx + d$.

The curve with equation $y = f(x)$ has gradient 4 at the point with coordinates $(0, -5)$.

(a) Find the values of c and d. [4]

The remainder when $f(x)$ is divided by $(x + 2)$ is 15 and the remainder when $f(x)$ is divided by $(x - 1)$ is 12.

(b) Find the values of a and b. [6]

2.3 Sequences and series

1 The third term of a geometric series is 15 and the common ratio of the series is 2.

(a) Find the sixth term of the series. [2]

(b) Find the sum of the first ten terms of the series. [3]

The second and fifth terms of the series form the first two terms of an arithmetic series.

(c) Find the ninth term of the arithmetic series. [5]

(d) Find the sum of the first thirteen terms of the arithmetic series. [3]

Answers on pages 39–57 **Answers** on pages 39–57 **Answers** on pages 39–57

2 The first, second and third terms of a geometric series are $(x + 4)$, $(x + 1)$ and x respectively.

(a) Find the value of x. [5]

(b) Find the common ratio of the series. [2]

(c) Find the sum to infinity of the series. [2]

3 Find $\sum_{r=1}^{10} (r + 3^r)$. [6]

4 A man invests £2000 at the beginning of each year in an account which pays 5% per annum compound interest.

(a) Find the amount in the account at the end of the third year. [2]

(b) Find the amount in the account at the end of n years. [5]

5 Find, to 3 significant figures, the value of $\sum_{r=6}^{12} (60 \times 0.8^r)$. [5]

6 The first three terms of a geometric series are $(3k + 1)$, $(2k - 1)$ and $(k + 1)$ respectively, where k is a constant and $k \neq 0$.

(a) Find the value of k. [5]

(b) Find the sum to infinity of the series. [4]

7 The third and fourth terms of a geometric series are 24 and 16 respectively.

(a) Find the first term and the common ratio of the series. [4]

(b) Find the sum to infinity of the series. [2]

8 Find the coefficient of x^3 in the expansion of

(a) $(1 + \frac{3}{4}x)^6$ [3]

(b) $(1 + \frac{1}{2}x)(1 + \frac{3}{4}x)^6$. [4]

9 **(a)** Expand $\left(\frac{2}{x} + \frac{x}{2}\right)^4$ in ascending powers of x. [4]

(b) Find the coefficient of x^6 in the expansion of $\left(\frac{2}{x} + \frac{x}{2}\right)^{10}$. [3]

10 **(a)** Find an expression in terms of n for the coefficient of x^2 in the expansion of $(1 + \frac{2}{3}x)^n$, where n is a positive integer. [2]

(b) Given that the coefficient of x^2 is 68, find the value of n. [4]

Answers on pages 39–57 **Answers** on pages 39–57 **Answers** on pages 39–57

2.4 Trigonometry

1 Solve, for $0° \leqslant x < 360°$, the equation $3\cos^2 x - 2\sin x = 2$. [7]

2 The curve with equation $y = 3 + k\sin x$ passes through the point with coordinates $\left(\frac{\pi}{2}, -1\right)$. Find

(a) the value of k [2]

(b) the greatest value of y. [2]

3 Given that $0 \leqslant x \leqslant \pi$, find the values of x for which

(a) $\cos 3x = 0.5$ [4]

(b) $\tan\left(x + \frac{\pi}{2}\right) = -1$. [3]

4 **(a)** Given that $\sin 15° = \frac{(\sqrt{6} - \sqrt{2})}{4}$, find in the form $\sqrt{m + n\sqrt{3}}$, where m and n are rational, the value of

(i) $\cos 15°$ **(ii)** $\sin 105°$. [6]

(b) Find, in radians to two decimal places, the values of x in the interval $0 \leqslant x \leqslant 2\pi$, for which

$$3\cos^2 x + \cos x - 2 = 0.$$ [6]

5 In the diagram, O is the centre of the circle and AB is a chord. The radius of the circle is 8 cm and $A\hat{O}B = 120°$.

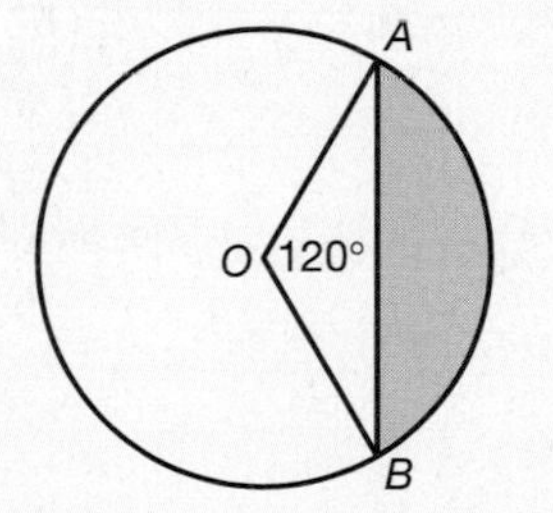

Calculate

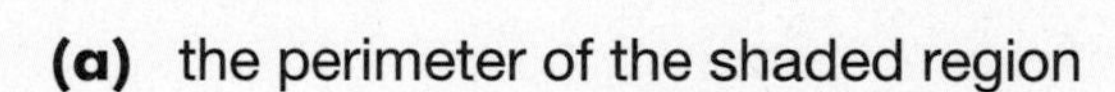

(a) the perimeter of the shaded region [6]

(b) the area of the shaded region. [5]

6 Find the values of θ in the interval $0° \leqslant \theta \leqslant 360°$ for which

(a) $\sin(\theta - 20°) = \frac{1}{\sqrt{2}}$ [4]

(b) $4\sin^2\theta = 1$. [5]

7 $f(x) = \sin\left(2x - \frac{\pi}{4}\right)$

(a) State the value of $f(0)$. [1]

(b) Solve the equation $f(x) = 0$, for $0 \leqslant x \leqslant 2\pi$, giving your answers in terms of π. [5]

(c) Sketch the curve with equation $y = f(x)$ for $0 \leqslant x \leqslant 2\pi$. [3]

Answers on pages 39–57 **Answers** on pages 39–57 **Answers** on pages 39–57

8

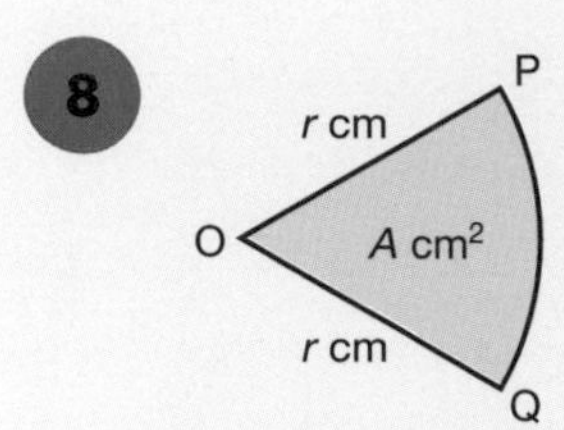

The figure shows a sector of a circle of radius r cm. The perimeter of the sector is 20 cm.

(a) Show that the area of the sector, A cm^2, is given by

$$A = (10r - r^2).$$ [4]

(b) If r is allowed to vary, show that the maximum area of the sector is 25 cm^2. [5]

(c) Find the range of values of r for which the area is more than 16 cm^2. [4]

9 Solve for values of x in the range $0 \leqslant x \leqslant 2\pi$, the equation

$$(\sin x - \cos x)^2 + \sin x = 1$$ [8]

10 Solve for values of x in the range $0 \leqslant x \leqslant 360°$, the equation

$$4\tan x + \frac{6}{\tan x} = 11,$$

giving your answers to 1 decimal place. [8]

Core 2

2.5 Exponentials and logarithms

1 By putting $y = 3^x$, or otherwise, solve the equation

$$3^{2x} = 3^{x+2} - 18.$$ [6]

2 Find, to 3 s.f., the values of x which satisfy the following:

(a) $e^{x-2} = 8$ [2]

(b) $\log_x 15 = 3$ [3]

(c) $2 \ln (2x - 1) = 5.$ [4]

3 Solve the following equation

$$3^x = 5^{2x-3}.$$ [4]

4 Find the value of

$$\sum_{r=1}^{20} \ln [5(3^r)],$$

expressing your answer in the form $a \ln 3 + b \ln 5$, where a and b are constants to be found. [5]

Answers on pages 39–57 **Answers** on pages 39–57 **Answers** on pages 39–57

5 Solve the following equations, giving your answers to 2 d.p.

(a) $\log_{10}(x + 1) - \log_{10}(x - 1) = 1$ [4]

(b) $2 + \ln x = \ln(x + 3)$. [4]

6 A curve C has equation

$$y = 10 - e^{2x}.$$

The points $A(\ln 2, a)$ and $B(b, -6)$ lie on C.

(a) Find the values of a and b. [5]

(b) Find the coordinates of the mid-point of AB. [2]

2.6 Differentiation

1 Find the coordinates of the turning points on the curve whose equation is

$$y = x^3 - 9x^2 + 24x.$$ [11]

2 A rectangular tank is made of thin sheet metal. The tank has a horizontal square base, of side x cm, and no top. When full the tank holds 500 litres.

(a) Show that the area, A cm^2, of sheet metal needed to make this tank is given by

$$A = x^2 + \frac{2\,000\,000}{x}, x \neq 0.$$ [6]

(b) Find the value of x which makes A a minimum and find this minimum value of A. [6]

(c) Prove that this value of A is a minimum. [4]

3 The function f is given by

$$f(x) = x + \frac{1}{4x}, x \neq 0.$$

(a) Find the values of x for which $f(x) = -\frac{5}{4}$. [4]

(b) Find the range of values of x for which f is an increasing function of x. [7]

4 **(a)** Find the coordinates of the stationary points on the curve with equation

$$y = x^3 + x^2 - x + 1.$$ [5]

(b) State which of your points are minimum points and which are maximum points, justifying your answers fully. [3]

(c) Find the range of values of x for which y is an increasing function of x. [4]

(d) Sketch the graph of y against x. [3]

5 $f(x) = 4x + 15 + \dfrac{9}{x}, x \neq 0$

(a) Find the coordinates of the points where the graph with equation $y = f(x)$ intersects the x-axis. [4]

(b) Find the coordinates of the turning points on the curve. [5]

6

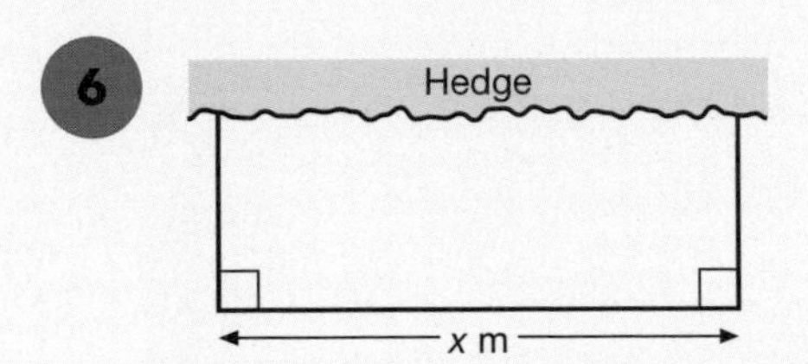

A farmer wishes to form a rectangular enclosure using an existing hedge as one of the sides of the enclosure. The length of the hedge side is x m, as shown in the figure.

The farmer has 200 m of fencing for the other three sides.

(a) Show that the area, A m^2, of the enclosure is given by

$$A = \tfrac{1}{2}(200x - x^2).$$ [4]

(b) Find the maximum possible area of the enclosure, justifying your answer fully. [7]

7 $f(x) = 2x - 3 + \dfrac{8}{x}, x \neq 0$

Find the range of values of x for which $f(x)$ is decreasing. [5]

8 $f(x) = 3x^{\frac{5}{2}} - 5x^{\frac{3}{2}}, x > 0$

(a) Show that $f'(x)$ can be written in the form $kx^{\frac{1}{2}}(x - 1)$ where k is a constant to be found. [3]

(b) Hence, or otherwise, find the coordinates of the turning point on the curve with equation $y = f(x)$. [3]

(c) Determine the nature of this point. [3]

Answers on pages 39–57 **Answers** on pages 39–57 **Answers** on pages 39–57

2.7 Integration

1

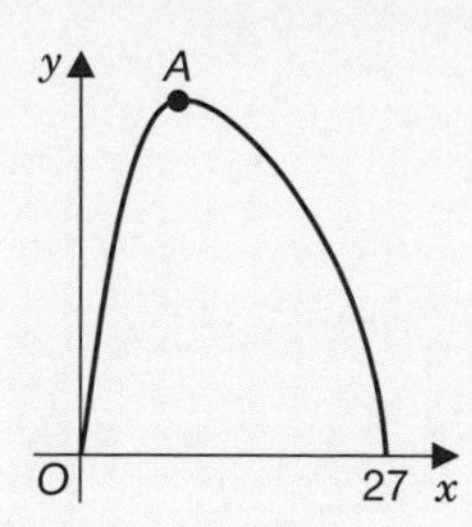

The diagram shows a sketch of the curve whose equation is

$y = 27x^{0.5} - x^{1.5}$ for $0 \leqslant x \leqslant 27$.

(a) Show that $\dfrac{dy}{dx} = 1.5x^{-0.5}(9 - x)$. [2]

The curve has a turning point at the point *A*.

(b) Find the coordinates of *A*. [3]

(c) Find the area of the finite region bounded by the curve and the x-axis. [5]

2

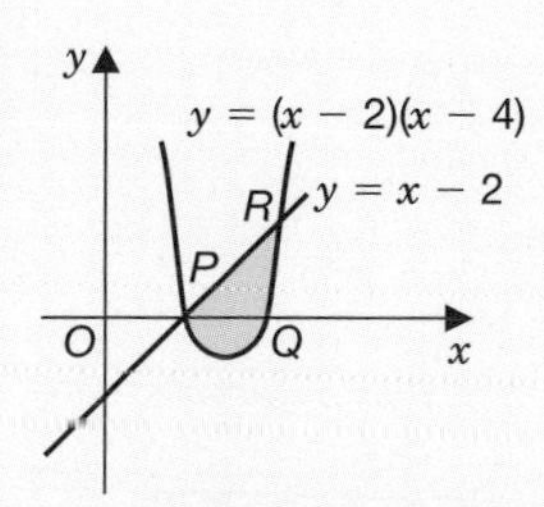

The diagram shows the line with equation $y = x - 2$ meeting the curve with equation $y = (x - 2)(x - 4)$ at the points *P* and *R*.

(a) Write down the coordinates of *P* and *Q*. [2]

(b) Find the coordinates of the point *R*. [4]

(c) Find the area of the shaded region bounded by the line and the curve. [6]

3

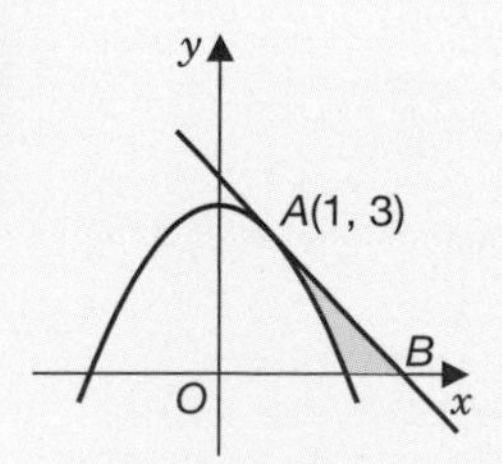

The tangent to the curve with equation $y = 4 - x^2$ at A(1, 3) meets the y-axis, the x-axis at the point B, as shown in the diagram.

(a) Find the x-coordinate of *B*. [7]

(b) Find the area of the shaded region. [9]

Answers on pages 39–57 **Answers** on pages 39–57 **Answers** on pages 39–57

4 $y = x^{\frac{3}{2}} - 3$

(a) Express y^2 in ascending powers of x. [2]

(b) Find $\int_1^4 y^2 \mathrm{d}x$. [2]

5 **(a)** Sketch the curve $y = (x - 3)^2$ for $0 \leqslant x \leqslant 5$. [2]

(b) Find, by integration, the area enclosed between the y-axis, the x-axis, the line $x = 1$ and part of the curve. [4]

6

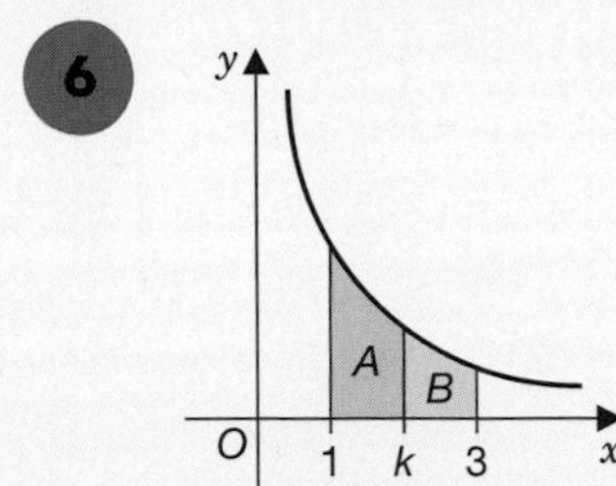

The diagram shows part of the curve with equation $y = \dfrac{1}{x^2}$. Find the value of k for which area A has the same value as area B. [6]

7 $y = 3x^{\frac{5}{2}} - 5x^{\frac{3}{2}}$, $x > 0$.

(a) Find the values of x for which $y = 0$. [4]

(b) Show that $\int_1^4 y\,dx = 46\frac{6}{7}$. [4]

8 **(a)** By using the substitution $p = x^{\frac{1}{2}}$, or otherwise, find the values of x for which

$$x - 3x^{\frac{1}{2}} + 2 = 0.$$ [4]

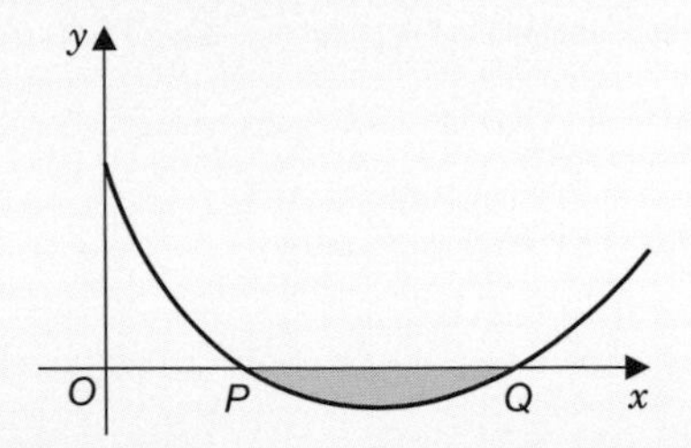

The diagram shows part of the graph of the curve with equation $y = x - 3x^{\frac{1}{2}} + 2$. The graph cuts the x-axis at the points P and Q.

(b) State the coordinates of P and Q. [1]

(c) Find, using integration, the area of the shaded region enclosed between the curve and the x-axis. [5]

9 Given that $f(x) = x(x^2 - 2)(3x - x^{-1})$

(a) express $f(x)$ in the form $px^4 + qx^2 + r$. [3]

(b) Evaluate $\int_0^2 f(x)\mathrm{d}x$. [5]

Answers on pages 39–57 **Answers** on pages 39–57 **Answers** on pages 39–57

Answers

2.1 Algebra and functions

examiner's tips

(1) (a) $(a+b)^4 = a^4 + 4a^3b + 6a^2b^2 + 4ab^3 + b^4$.

> **In each term the sum of the powers of a and b must be 4; the coefficients come from Pascal's Triangle.**

(b) Putting $a = x$ and $b = (1-x)$ on the RHS gives the expression requiring simplification. Hence, from part **(a)**, it simplifies to $(x+1-x)^4 = 1^4 = 1$.

> **A much longer method would be to expand all the brackets and then collect terms! It is always worth a few seconds thought before launching into a routine method.**

(2) $15 = a4^n$ — **Since both points lie on the curve.**

$9 = a2^n$

dividing gives — **To eliminate a.**

$\frac{5}{3} = 2^n$ — **Note that $4^n/2^n = (\frac{4}{2})^n = 2^n$**

$\log \frac{5}{3} = \log 2^n$ — **Generally if the unknown is a power we take logs, base 10, of both sides.**

$\log \frac{5}{3} = n\log 2$

$\frac{\log \frac{5}{3}}{\log 2} = n$, i.e. $n = 0.737$ (3 s.f.)

hence, $9 = \frac{5a}{3}$ — **Since $\frac{5}{3} = 2^n$.**

$a = \frac{27}{5} = 5.4$

(3) (a) $\ln(a+10) = 2\ln(a-2)$ — **Note $a > 2$ as we cannot take the log of a negative number.**

$\ln(a+10) = \ln(a-2)^2$ — **Using the 3rd law of logarithms.**

$(a+10) = (a-2)^2 = a^2 - 4a + 4$

$a^2 - 5a - 6 = 0$

$(a-6)(a+1) = 0$

$a = 6$ or $a = -1$

but $a > 2$, so $a = 6$.

(b) $2^{2x} - 6.2^x + 8 = 0$ — **Taking logs here doesn't help since log 0 is undefined and we cannot simplify the logarithm of a sum or difference.**

$(2^x)^2 - 6.2^x + 8 = 0$ — **The key step – we now have a quadratic in 2^x.**

$(2^x - 4)(2^x - 2) = 0$

$2^x = 4$ or $2^x = 2$

$x = 2$ or $x = 1$

(4) (a) $f(1) = 2 + p - 1 + q = 0 \rightarrow p + q = -1$ — **Using the Factor Theorem.**

$f(-2) = -16 + 4p + 2 + q = 0 \rightarrow 4p + q = 14$ — **Using the Factor Theorem.**

subtracting gives $3p = 15$, $p = 5$ and $q = -6$.

(b) $f(x) = (x-1)(x+2)(2x+3)$ — **By inspection of the constant terms.**

Core 2

(5) (a) $f(-2) = 15$ ← Using the Remainder Theorem.

$-8p + 44 - 4p - 5 = 15$ ← Putting $x = -2$.

$12p = 24$

$p = 2$

(b) $f(x) = 2x^3 + 11x^2 + 4x - 5$ ← Now look for a value of x for which $f(x) = 0$.

$f(1) = 2 + 11 + 4 - 5 \neq 0$ ← Try $x = 1, -1, 2, -2$ etc.

$f(-1) = -2 + 11 - 4 - 5 = 0 \rightarrow (x + 1)$ is a factor of $f(x)$ ← By the Factor Theorem.

$\rightarrow f(x) = (x + 1)(2x^2 + 9x - 5)$ ← By inspection – the first and third terms are easy!

$= (x + 1)(x + 5)(2x - 1)$

(c) $(x + 1)(x + 5)(2x - 1) = (x + 1)(x + 5)$ ← Don't multiply out!

$(x + 1)(x + 5)(2x - 1) - (x + 1)(x + 5) = 0$ ← Collecting the terms.

$(x + 1)(x + 5)\{(2x - 1) - 1\} = 0$ ← Taking out the common factor.

$(x + 1)(x + 5)(2x - 2) = 0$

$x = -1$ or -5 or 1

(6) Let $P(x) = (x - a)^2Q(x)$ ← Since $P(x)$ is divisible by $(x - a)^2$.

$P'(x) = 2(x - a)Q(x) + (x - a)^2Q'(x)$ ← Using the Product Rule.

$= (x - a)\{2Q(x) + (x - a)Q'(x)\}$ ← Taking out the common factor.

Hence $P'(x)$ is divisible by $(x - a)$.

Let $f(x) = x^4 + x^3 - 12x^2 + px + q$

Then $f'(x) = 4x^3 + 3x^2 - 24x + p$ is divisible by $(x + 2)$ ← Using the above result.

So $f'(-2) = 0$ i.e. $-32 + 12 + 48 + p = 0$ ← Using the Factor Theorem.

i.e. $p = -28$

Also $f(-2) = 0$ i.e. $16 - 8 - 48 + 56 + q = 0$ ← Using the Factor Theorem.

i.e. $q = -16$

(7) (a) $P(x) = ax^3 - 3x^2 + bx + 6$ ← Two unknowns, so we need 2 equations.

$P(1) = a - 3 + b + 6 = -6$ ← Using the remainder theorem.

$\Rightarrow a + b = -9$ (1)

$P(-2) = -8a - 12 - 2b + 6 = 0$ ← Using the factor theorem.

$\Rightarrow -8a - 2b = 6$

$\Rightarrow 4a + b = -3$ (2) ← Dividing by -2; we now have two simultaneous equations.

$3a = 6$

$a = 2 \Rightarrow b = -11$ ← Substituting back to find b.

(b) $P(x) = 0 \Rightarrow 2x^3 - 3x^2 - 11x + 6 = 0$

$(x + 2)(2x^2 - 7x + 3) = 0$ ← $(x + 2)$ is a factor from above.

$(x + 2)(2x - 1)(x - 3) = 0$ ← Factorising the quadratic.

$\Rightarrow x = -2$ or $\frac{1}{2}$ or 3 ← Equating each factor to 0 and solving.

Core 2

(8) (a) Let $f(x) = 3x^3 + x^2 + px + 24$

then $f(-4) = -192 + 16 - 4p + 24 = 0$ ← **Using the factor theorem.**

$\Rightarrow \quad 4p = -152$

$\Rightarrow \quad p = -38$

(b) Let $P(x) = 5x^3 - 3x^2 + ax + 7$ ← **There are two unknowns, so we need two equations.**

$P(-2) = -40 - 12 - 2a + 7 = r$ ← **Using the remainder theorem.**

$\Rightarrow \quad -45 = r + 2a$ ← **Rearrange to put unknowns together.**

Let $Q(x) = 4x^3 + ax^2 + 7x - 4$

$Q(-2) = -32 + 4a - 14 - 4 = 2r$ ← **Using the remainder theorem.**

$\Rightarrow \quad -50 = 2r - 4a$

$-25 = r - 2a$ ← **Dividing by 2.**

From above, $-45 = r + 2a$ ← **Subtract to eliminate r.**

$20 = -4a$

$-5 = a$

(9) (a) $f(x) = ax^3 + bx^2 + cx + d$

$f(0) = d = -5$ ← **Since $(0, -5)$ lies on the curve.**

$f'(x) = 3ax^2 + 2bx + c$

$f'(0) = c = 4$ ← **Since curve has gradient 4 at $(0, -5)$.**

(b) $f(1) = a + b + 4 - 5 = 12$ ← **Using the reminder theorem.**

$\Rightarrow \quad a + b = 13 \quad (1)$

$f(-2) = -8a + 4b - 8 - 5 = 15$ ← **By the remainder theorem again.**

$\Rightarrow \quad -8a + 4b = 28$

$\Rightarrow \quad -2a + b = 7 \quad (2)$ ← **Dividing by 4 to simplify.**

$3a = 6$ ← **Subtracting (2) from (1).**

$a = 2$

$\Rightarrow \quad b = 11$

(10) $g(x) = 2x^3 - x^2 - 23x - 20$

(a) $g(-1) = -2 - 1 + 23 - 20 = 0$

$\Rightarrow (x + 1)$ is a factor of $g(x)$ ← **Using the factor theorem.**

(b) $g(x) = (x + 1)(2x^2 - 3x - 20)$ ← **Factorising by inspection.**

$= (x + 1)(2x + 5)(x - 4)$ ← **Factorising the quadratic.**

(c) $g(x) = 0 \Rightarrow (x + 1)(2x + 5)(x - 4) = 0$

$\Rightarrow x = -1$ or $-\frac{5}{2}$ or 4 ← **Equating each factor to 0.**

2.2 Coordinate geometry

(1) (a) $x^2 + y^2 - 16x - 12y + 96 = 0$

$x^2 - 16x + 64 + y^2 - 12y + 36 = 64 + 36 - 96$ ← **Completing the square twice.**

$(x - 8)^2 + (y - 6)^2 = 4$ ← **Compare with the standard form $(x - a)^2 + (y - b)^2 = r^2$**

Centre is (8, 6); radius is 2

(b)

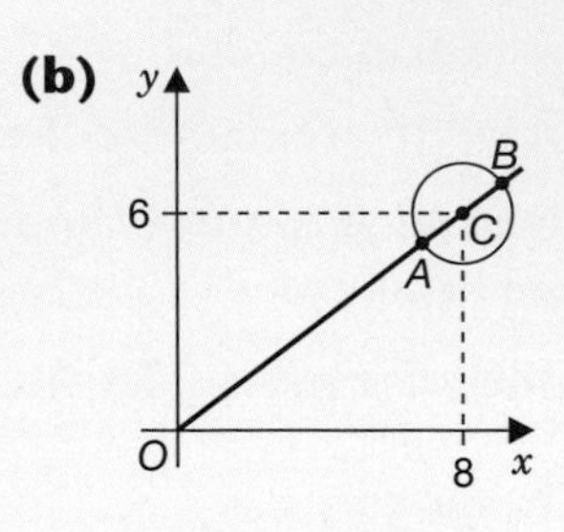

$OC = \sqrt{6^2 + 8^2} = 10$ ← **Drawing a diagram is essential.**

OA = least distance ← **A is the point closest to O.**

$= 10 - 2$

$= 8$

OB = greatest distance ← **B is the point furthest away from O.**

$= 10 + 2$

$= 12$

(2) Let A be (0, 4); B be (0, −4) and C be (3, 5). ← **The centre of the circle must lie on the perpendicular bisector of any chord of the circle.**

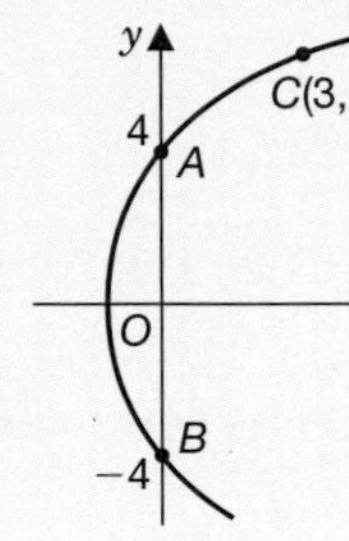

AB: Centre lies on Ox

AC: Mid-point is $\left(\frac{3}{2}, \frac{9}{2}\right)$ ← **Using $\left(\frac{x_1 + x_2}{2}, \frac{y_1 + y_2}{2}\right)$.**

Gradient of AC $= \frac{5 - 4}{3 - 0} = \frac{1}{3}$ ← **Using $\frac{y_2 - y_1}{x_2 - x_1}$.**

$\therefore$ Gradient of perpendicular $= -3$ ← **Using $m_1 m_2 = -1$.**

$\therefore$ Equation is $y - \frac{9}{2} = -3\left(x - \frac{3}{2}\right)$ ← **Using $y - y_1 = m(x - m_1)$.**

$2y - 9 = -6x + 9$

$2y = -6x + 18$

Now find where this line meets the x-axis, $y = 0$.

$0 = -6x + 18$

$3 = x$ i.e. centre is (3, 0)

so, radius is 5 ← **Since (3, 5) is on circle.**

Hence, equation of circle is

$(x - 3)^2 + y^2 = 5^2$ ← **Using $(x - a)^2 + (y - b)^2 = r^2$.**

$\Rightarrow x^2 - 6x + 9 + y^2 = 25$ ← **Multiplying out.**

$\Rightarrow x^2 + y^2 - 6x - 16 = 0$ ← **Collecting terms.**

(3) (a) $(x - 1)^2 + (y - 4)^2 = 3^2$ ← **Using $(x - a)^2 + (y - b)^2 = r^2$.**

$x^2 + y^2 - 2x - 8y + 8 = 0$

(b) $(2.8 - 1)^2 + (1.6 - 4)^2$ ← **Substitute the coordinates into the L.H.S. of the equation.**

$= 1.8^2 + (-2.4)^2 = 3^2$

The point lies on the circle.

Core 2

(4) (a) $x^2 + y^2 - 8x - 10y - 59 = 0$

$x^2 - 8x + 16 + y^2 - 10y + 25 = 59 + 16 + 25$ ← **Completing the square.**

$(x - 4)^2 + (y - 5)^2 = 10^2$ ← **Compare with $(x - a)^2 + (y - b)^2 = r^2$.**

so, centre is (4, 5); radius is 10.

(b) $(10 - 4)^2 + (13 - 5)^2$

$= 36 + 64$

$= 100$

Hence (10, 13) lies on circle.

(c) Gradient of radius $OP = \frac{13 - 5}{10 - 4} = \frac{4}{3}$ ← **Using $\frac{y_1 - y_2}{x_1 - x_2}$ = gradient.**

$\therefore$ Gradient of tangent at $P = -\frac{3}{4}$ ← **Flip the fraction and change sign.**

$\therefore$ Equation of tangent at P is

$y - 13 = -\frac{3}{4}(x - 10)$ ← **Using $y - y_1 = m(x - x_1)$.**

$4y - 52 = -3x + 30$

$4y + 3x - 82 = 0$

(5) (a) $x^2 + y^2 - 8x - 24y - 9 = 0$

$x^2 - 8x + 16 + y^2 + 24y + 144 = 16 + 144 + 9$ ← **Completing the square twice.**

$(x - 4)^2 + (y + 12)^2 = 13^2$ ← **Compare with $(x - a)^2 + (y - b)^2 = r^2$.**

Centre is (4, −12); radius is 13.

(b) $x + 1 = 0 \Rightarrow x = -1$

$1 + y^2 + 8 + 24y - 9 = 0$ ← **Substituting in.**

$y^2 + 24y = 0$

$y(y + 24) = 0$ ← **Factorise – do not cancel y.**

$y = 0$ or $y = -24$

$\therefore$ points are (−1, 0) and (−1, −24) ← **State the coordinates of each point.**

(c) By inspection, $AB = 24$

(d) Gradient of radius $= \frac{0 - -12}{-1 - 4}$ ← **Using $\frac{y_1 - y_2}{x_1 - x_2}$.**

$= -\frac{12}{5}$

$\Rightarrow$ Gradient of tangent $= +\frac{5}{12}$ ← **Flip fraction and change the sign.**

$\Rightarrow$ Equation of tangent is $y - 0 = \frac{5}{12}(x - -1)$ ← **Using $y - y_1 = m(x - x_1)$.**

$\Rightarrow \quad y = \frac{5}{12}(x + 1)$

$12y - 5x - 5 = 0$

(e) Put $y = -12$: $-12 = \frac{5}{12}(x + 1)$ ← **Solving simultaneously.**

$\Rightarrow \quad -\frac{144}{5} = x + 1$

$\Rightarrow \quad -\frac{149}{5} = x$

Core 2

(6) (a) $x^2 + y^2 + 10x + 20y + 25 = 0$

$x^2 + 10x + 25 + y^2 + 20y + 100 = 25 + 100 - 25$ ← **Completing the square twice.**

$(x + 5)^2 + (y + 10)^2 = 10^2$ ← **Comparing with $(x - a)^2 + (y - b)^2 = r^2$.**

Centre is $(-5, -10)$; radius is 10.

$x^2 + y^2 - 8x - 4y - 5 = 0$

$x^2 - 8x + 16 + y^2 - 4y + 4 = 16 + 4 + 5$ ← **Completing the square twice.**

$(x - 4)^2 + (y - 2)^2 = 5^2$

Centre is $(4, 2)$; radius is 5. ← **Comparing with $(x - a)^2 + (y - b)^2 = r^2$.**

(b) Distance between centres

$= \sqrt{(4 - -5)^2 + (2 - -10)^2}$ ← **Using $\sqrt{(x_1 - x_2)^2 + (y_1 - y_2)^2}$.**

$= \sqrt{9^2 + 12^2}$

$= 15$

Sum of radii $= 5 + 10 = 15$ ←

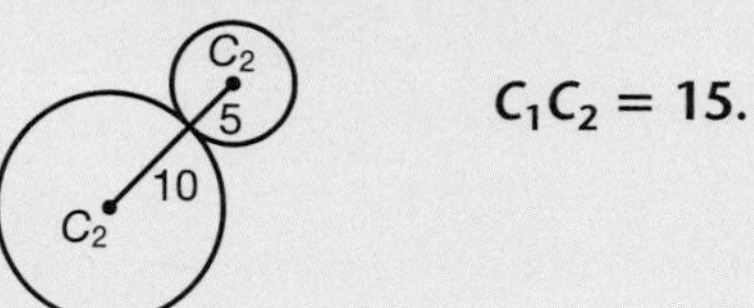

$C_1C_2 = 15$.

Hence circles must touch.

(7) radius $= \sqrt{(2 - 5)^2 + (-3 - -7)^2}$ ← **Using $\sqrt{(x_1 - x_2)^2 + (y_1 - y_2)^2}$.**

$= \sqrt{3^2 + 4^2}$

$= 5$ ← **3, 4, 5 △.**

equation is $(x - 2)^2 + (y - -3)^2 = 5^2$ ← **Using $(x - a)^2 + (y - b)^2 = r^2$.**

$x^2 + y^2 - 4x + 6y + 4 + 9 - 25 = 0$ ← **Multiplying out.**

$x^2 + y^2 - 4x + 6y - 12 = 0$ ← **Collecting the terms**

(8) $x^2 + (mx)^2 - 6x - 6(mx) + 17 = 0$ ← **Solving the two equations simultaneously.**

$(1 + m^2)x^2 - 6(1 + m)x + 17 = 0$ ← **If $y = mx$ is a tangent this equation will have repeated roots so '$b^2 = 4ac$'.**

$\Rightarrow$ $36(1 + m)^2 = 68(1 + m^2)$

$32m^2 - 72m + 32 = 0$ ← **Collecting terms; this equation does not factorise so use the quadratic formula.**

$\Rightarrow$ $4m^2 - 9m + 4 = 0$

$\Rightarrow$ $m = \dfrac{9 \pm \sqrt{17}}{8}$.

(9) (a) $f(0) = -5 \Rightarrow d = -5$

$f'(x) = 3ax^2 + 2bx + c$ ← **First find the derivative of $f(x)$.**

$f'(0) = 4 \Rightarrow c = 4$

(b) $f(-2) = 15$ ← **Using the Remainder Theorem.**

$-8a + 4b - 8 - 5 = 15$ i.e. $-2a + b = 7$

$f(1) = 12$ ← **Using the Remainder Theorem.**

$a + b + 4 - 5 = 12$ i.e. $a + b = 13$

$3a = 6 \Rightarrow a = 2$ ← **Subtracting the equations to eliminate b.**

$\Rightarrow b = 11$

Core 2

2.3 Sequences and series

(1) (a) $a(2)^2 = 15 \rightarrow 4a = 15 \rightarrow a = \frac{15}{4}$

6th term $= \frac{15}{4}.\ 2^5 = \frac{15}{4}.\ 32 = 120$ ← **nth term $= ar^{n-1}$ – you need to learn this!**

(b) $S_{10} = \frac{15}{4}(2^{10} - 1)/(2 - 1)$ ← **Use the form $Sn = a(r^n - 1)/(r - 1)$, since $r > 1$.**

$= 3836.25$

(c) 2nd term $= \frac{15}{4} \times 2 = 7.5$; 5th term $= \frac{15}{4} \times 2^4 = 60$

hence first and second terms of *AP* are 7.5 and 60

i.e. $a = 7.5$ and $d = 52.5$ ← **nth term $= a + (n - 1)d$ – you need to learn this!**

so 9th term $= a + 8d = 7.5 + 8 \times 52.5$

$= 427.5$

(d) 13th term $= a + 12d = 7.5 + 12 \times 52.5 = 637.5$

$S_{13} = \frac{(7.5 + 637.5)}{2} \times 13$ ← **Here we're using $Sn = \frac{(a + l)n}{2}$.**

$= 4192.5$

(2) (a) $\frac{(x + 1)}{(x + 4)} = \frac{x}{(x + 1)}$ ← **Since the ratios of successive terms are equal.**

$\Rightarrow \quad (x + 1)^2 = x(x + 4)$ ← **Cross-multiplying.**

$\Rightarrow x^2 + 2x + 1 = x^2 + 4x$

$\Rightarrow \quad 2x + 1 = 4x$

$\Rightarrow \quad x = \frac{1}{2}$

(b) $r = \frac{x}{(x + 1)} = \frac{1}{2}/(\frac{1}{2} + 1) = \frac{1}{3}$

(c) $S_\infty = \frac{a}{(1 - r)}$ ← **You need to learn this result for the sum to infinity of a GP but note that it only applies for $-1 < r < 1$.**

and $\quad a = x + 4 = 4.5$

So, $S_\infty \ = 4.5/(1 - \frac{1}{3})$

$= 6.75$

(3) $\sum_{r=1}^{10} (r + 3^r) = \sum_{r=1}^{10} r + \sum_{r=1}^{10} 3^r$ ← **The first series is an AP, with $a = 1$ and $d = 1$ and $n = 10$. The second series is a GP, with $a = 3$ and $r = 3$ and $n = 10$.**

$= \frac{(1 + 10)}{2} \times 10 + 3\frac{(3^{10} - 1)}{(3 - 1)}$

$= 55 + 88\,572 = 88\,627$

(4) (a)

End of Year	*Amount*	
1	2000×1.05	← **To increase by 5%, multiply by 1.05.**
2	$(2000 + 2000 \times 1.05) \times 1.05$	← **£2000 is added at the start of the year.**
	$= (2000 \times 1.05) + (2000 \times 1.05^2)$	← **Multiply out.**
	$= 2000\,(1.05 + 1.05^2)$	← **Factorise.**

so, at end of Year 3, the amount will be:

$2000(1.05 + 1.05^2 + 1.05^3)$ ← **Following the pattern for Years 1 and 2.**

$= £6620.25$

(b) At end of Year n,

$2000(1.05 + 1.05^2 + \dots + 1.05^n)$ ← This is a geometric series with common ratio, $r = 1.05$.

$= 2000 \times 1.05 \dfrac{(1.05^n - 1)}{(1.05 - 1)}$ ← Using $S_n = a\dfrac{(r^n - 1)}{(r - 1)}$.

$= £42\,000\,(1.05^n - 1)$ ← Check by putting $n = 3$ for answer to **(a)**.

(5) $\sum_{r=6}^{12} (60 \times 0.8^r)$

$= 60 \times 0.8^6 + 60 \times 0.8^7 + \dots + 60 \times 0.8^{12}$ ← Write out the series in full to help to decide what type it is.

$a = 60 \times 0.8^6;\ r = 0.8$

$S_7 = 60 \times 0.8^6 \dfrac{(0.8^7 - 1)}{(0.8 - 1)}$ ← There are 7 terms: use $a\dfrac{(r^n - 1)}{(r - 1)}$.

$= 62.15 \dots = 62.2$ (3 s.f.) ← 3 s.f. required.

(6) (a) $(3k + 1) + (2k - 1) + (k + 1) + \dots$

$\Rightarrow \dfrac{(2k - 1)}{(3k + 1)} = \dfrac{(k + 1)}{(2k - 1)} (= r)$ ← Successive terms are in the same ratio.

$\Rightarrow (2k - 1)^2 = (3k + 1)(k + 1)$ ← Cross-multiplying.

$\Rightarrow 4k^2 - 4k + 1 = 3k^2 + 4k + 1$ ← Multiplying out.

$\Rightarrow k^2 - 8k = 0$ ← Collecting terms.

$\Rightarrow k(k - 8) = 0$ ← Factorising.

$\Rightarrow k = 8$ since $k \neq 0$

(b) The first term is $(3 \times 8) + 1 = 25$

The common ratio is $\left(\dfrac{8 + 1}{16 - 1}\right) = \dfrac{9}{15} = \dfrac{3}{5}$ ← Since $-1 < r < 1$, there is a sum to infinity.

$S_\infty = \dfrac{25}{1 - \frac{3}{5}} = \dfrac{125}{5 - 3} = 62.5$ ← Using $S_\infty = \dfrac{a}{1 - r}$.

(7) (a) $ar^2 = 24$ (1) ← nth term of a geometric series is ar^{n-1}.

$ar^3 = 16$ (2)

$r = \frac{16}{24} = \frac{2}{3}$ ← Dividing (2) by (1).

$\Rightarrow a \times \frac{4}{9} = 24$ ← Substituting back into (1).

$\Rightarrow a = 54$

(b) $S_\infty = \dfrac{a}{1 - r} = \dfrac{54}{1 - \frac{2}{3}} = 162$

(8) (a) Term in $x^3 = {}^6_3C(\frac{3}{4}x)^3 \times 1^3$ ← Remember ${}^6_3C$ can be written $\binom{6}{3}$.

Coefficient of $x^3 = \dfrac{6 \times 5 \times 4}{1 \times 2 \times 3} \times \dfrac{3^3}{4^3}$ ← You can use a calculator to work out ${}^6_3C$.

$= \dfrac{20}{1} \times \dfrac{27}{64} = \dfrac{135}{16}$

(b) Coefficient of x^3 in $\left(1 + \dfrac{1}{2}x\right)\left(1 + \dfrac{3x}{4}\right)^6$

$= \dfrac{1}{2} \times$ coefficient of x^2 in $\left(1 + \dfrac{3x}{4}\right)^6$ + answer to **(a)**

$= \dfrac{1}{2} \times {}^6_2C\left(\dfrac{3}{4}\right)^2 + \dfrac{135}{16}$

$= \dfrac{1}{2} \times \dfrac{6 \times 5}{1 \times 2} \times \dfrac{9}{16} + \dfrac{135}{16} = \dfrac{405}{32}$

(9) (a) $\left(\frac{2}{x}+\frac{x}{2}\right)^4$

$= \left(\frac{2}{x}\right)^4 + \binom{4}{1}\left(\frac{2}{x}\right)^3\left(\frac{x}{2}\right)^1 + \binom{4}{2}\left(\frac{2}{x}\right)^2\left(\frac{x}{2}\right)^2 + \binom{4}{3}\left(\frac{2}{x}\right)^1\left(\frac{x}{2}\right)^3 + \left(\frac{x}{2}\right)^4$ ← Always write out <u>in full</u> first before simplifying.

$= \frac{16}{x^4} + 4 \times \frac{8}{x^3} \times \frac{x}{2} + 6 \times \frac{4}{x^2} \times \frac{x^2}{4} + 4 \times \frac{2}{x} \times \frac{x^3}{8} + \frac{x^4}{16}$

$= 16x^{-4} + 16x^{-2} + 6 + x^2 + \frac{x^4}{16}$ ← Cancelling and using negative powers.

(b) Term in $x^6 = {}^{10}_{2}\mathrm{C}\left(\frac{2}{x}\right)^2\left(\frac{x}{2}\right)^8$ ← Always write down the complete term first even if you only want the coefficient.

$= \frac{10 \times 9}{1 \times 2} \times \frac{4}{x^2} \times \frac{x^8}{256}$

$= \frac{45}{64}x^6$

Hence, coefficient is $\frac{45}{64}$.

(10) (a) Term in $x^2 = \binom{n}{2}\left(\frac{2}{3}x\right)^2$ ← Remember that $\binom{n}{r} = {}^n_r\mathrm{C}$ and ${}^n_r\mathrm{C} = \frac{n!}{r!(n-r)!}$ $= \frac{n(n-1)\ldots(n-r+1)}{r!}$

$= \frac{n(n-1)}{1 \times 2} \times \frac{4x^2}{9}$

Hence, coefficient $= \frac{2n(n-1)}{9}$

(b) $\frac{2n(n-1)}{9} = 68$

$\Rightarrow \quad n(n-1) = 306$ ← Since n is a positive integer and $306 = 18 \times 17$.

$\Rightarrow \quad n = 18$

2.4 Trigonometry

(1) $3(1 - \sin^2 x) - 2\sin x = 2$ ← Use $\sin^2 x + \cos^2 x = 1$ to give a quadratic in $\sin x$.

$3\sin^2 x + 2\sin x - 1 = 0$ ← Collect all the terms on one side so that the $\sin^2 x$ term has a positive coefficient.

$(3\sin x - 1)(\sin x + 1) = 0$ ← Factorise or use the quadratic formula.

$\sin x = \frac{1}{3}$ or $\sin x = -1$

$x = 19.5°, 160.5°$ or $x = 270°$ ← sin is positive in quadrants 1 and 2.

(2) (a) $-1 = 3 + k\sin\frac{\pi}{2}$ ← Substituting for x and y.

$\rightarrow k = -4$

(b) so $y = 3 - 4\sin x$

hence $y_{max} = 7$ ← Greatest value of y will occur when $\sin x = -1$.

(3) (a) $\cos 3x = 0.5$ where $0 \leqslant 3x \leqslant 3\pi$ ← Change the range so that we know how many values to write down.

$3x = \frac{\pi}{3}, \frac{5\pi}{3}, \frac{7\pi}{3}$

$x = \frac{\pi}{9}, \frac{5\pi}{9}, \frac{7\pi}{9}$ ← Dividing by 3. Change the range.

(b) $\tan\left(x + \frac{\pi}{2}\right) = -1$ where $\frac{\pi}{2} \leqslant \left(x + \frac{\pi}{2}\right) \leqslant \frac{3\pi}{2}$

$x + \frac{\pi}{2} = \frac{3\pi}{4}, \frac{7\pi}{4}$

hence $x = \frac{\pi}{4}$ is the only solution in the given range.

(4) (a) (i) Draw a right-angled triangle as follows:

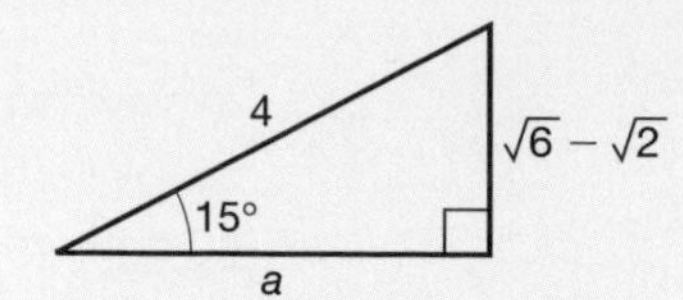

Now find the missing adjacent side by using Pythagoras:

$a^2 = 4^2 - (\sqrt{6} - \sqrt{2})^2 = 16 - (6 - 2\sqrt{6}\sqrt{2} + 2)$ ← **Multiplying out the brackets.**

$= 8 + 2\sqrt{12}$

$= 8 + 4\sqrt{3}$ ← **Since $\sqrt{12} = \sqrt{4 \times 3} = 2\sqrt{3}$.**

$a = \sqrt{(8 + 4\sqrt{3})}$ so

$\cos 15° = \dfrac{\sqrt{(8 + 4\sqrt{3})}}{4} = \sqrt{(8 + 4\sqrt{3})/16} = \sqrt{\tfrac{1}{2} + \tfrac{1\sqrt{3}}{4}}$

(ii) $\sin 105° = \sin 75° = \cos 15° = \sqrt{\tfrac{1}{2} + \tfrac{1\sqrt{3}}{4}}$ ← **Since $\sin x = \sin(180° - x)$ and $\cos 15° = \sin 75°$ from the triangle.**

(b) $3\cos^2 x + \cos x - 2 = 0$

$(3\cos x - 2)(\cos x + 1) = 0$ ← **Factorise or use the formula.**

$\cos x = \tfrac{2}{3}$ or $\cos x = -1$

$x = 0.84$ or $(2\pi - 0.84) = 5.44$ or $x = \pi = 3.14$ (2 d.p.)

i.e. $x = 0.84, 5.44$ or 3.14

(5) (a) $120° = \dfrac{2\pi}{3}$ ← **We need to use $s = r\theta$ where θ is in radians.**

Perimeter $= AB + \text{arc } AB$

$= 2 \times 8\sin 60° + 8 \times \dfrac{2\pi}{3}$ ← **ΔOAB is isosceles so the perpendicular from O to AB meets AB at its mid-point, N, say where $AB = 2AN = 2 \times 8\sin 60$.**

$= 8\sqrt{3} + 16\dfrac{\pi}{3} = 30.6$ cm (3 s.f.)

(b) Area $=$ Area of sector OAB − Area ΔOAB

$= \tfrac{1}{2}8^2\dfrac{2\pi}{3} - \tfrac{1}{2}8^2 \sin\dfrac{2\pi}{3}$ ← **Area of a sector is $\tfrac{1}{2}r^2\theta$ and area of Δ is $\tfrac{1}{2}ab\sin C$.**

$= 39.3 \text{ cm}^2$

(6) (a) $\sin(\theta - 20°) = \dfrac{1}{\sqrt{2}}$ $\quad 0° \leqslant \theta \leqslant 360°$

$-20° \leqslant \theta - 20° \leqslant 340°$ ← **Change the range.**

$\Rightarrow \theta - 20° = 45°$ or $135°$ ← **sine is positive in the first and second quadrant.**

$\Rightarrow \theta = 65°, 155°$

(b) $4\sin^2\theta = 1$

$\Rightarrow \sin^2\theta = \tfrac{1}{4}$

$\Rightarrow \sin\theta = \tfrac{1}{2}$ or $\sin\theta = -\tfrac{1}{2}$ ← **Two square roots.**

$\theta = 30°, 150°$ $\quad \theta = 210°, 330°$

(7) (a) $f(0) = \sin\left(-\frac{\pi}{4}\right)$

$= -\sin\frac{\pi}{4}$

$= -\frac{1}{\sqrt{2}}$

sin (−x) = −sin x. You should know the surd forms.

(b) $0 = \sin\left(2x - \frac{\pi}{4}\right)$

$0 \leqslant x \leqslant 2\pi \Rightarrow 0 \leqslant 2x \leqslant 4\pi$

$\Rightarrow -\frac{\pi}{4} \leqslant \left(2x - \frac{\pi}{4}\right) \leqslant \frac{15\pi}{4}$ — **Change the range.**

$2x - \frac{\pi}{4} = 0, \pi, 2\pi, 3\pi$ — **These are the angles in the range.**

$2x = \frac{\pi}{4}, \frac{5\pi}{4}, \frac{9\pi}{4}, \frac{13\pi}{4}$

$x = \frac{\pi}{8}, \frac{5\pi}{8}, \frac{9\pi}{8}, \frac{13\pi}{8}$ — **Rearrange to give the values of x.**

(8) (a) Let angle in sector be θ radians.

then, $2r + r\theta = 20$ — **Using $s = r\theta$ for the arc length.**

$\Rightarrow \quad \theta = \frac{20}{r} - 2$

$A = \frac{1}{2}r^2\theta$

$= \frac{1}{2}r^2\left(\frac{20}{r} - 2\right)$ — **Substituting for θ.**

i.e. $A = 10r - r^2$

(b) $\frac{dA}{dr} = 10 - 2r = 0 \Rightarrow r = 5$ — **At maximum, $\frac{dA}{dr} = 0$.**

$\frac{d^2A}{dr^2} = -2 < 0 \qquad \Rightarrow$ maximum

$A_{max} = (10 \times 5) - 5^2 = 25$ — **Substituting $r = 5$.**

(c) $A > 16$

$\Rightarrow \quad 10r - r^2 > 16$

$0 > r^2 - 10r + 16$

$0 > (r - 8)(r - 2)$ — **Factorising.**

$2 < r < 8$ — **You could check your answer by putting $r = 4$ say.**

(9) $(\sin x - \cos x)^2 + \sin x = 1$

$\sin^2 x + \cos^2 x - 2\sin x \cos x + \sin x = 1$ — **Multiplying out.**

$\sin x - 2\sin x \cos x = 0$ — **Since $\sin^2 x + \cos^2 x = 1$.**

$\sin x(1 - 2\cos x) = 0$ — **Factorise don't cancel.**

$\sin x = 0$ or $\cos x = \frac{1}{2}$

$x = 0, \pi, 2\pi \qquad x = \frac{\pi}{3}, \frac{5\pi}{3}$

(10) $4\tan x + \frac{6}{\tan x} = 11$

$4\tan^2 x + 6 = 11\tan x$ — **'Cleaning' the fraction.**

$4\tan^2 x - 11\tan x + 6 = 0$

$(4\tan x - 3)(\tan x - 2) = 0$ — **Factorising (or use formula).**

$\tan x = \frac{3}{4}$ or $\tan x = 2$

$x = 36.9°$ or $216.9° \qquad x = 63.4°$ or $243.4°$ — **Answers to 1 decimal place. The tan graph repeats every 180°.**

Core 2

2.5 Exponentials and logarithms

(1) $3^{2x} = 3^{x+2} - 18$

$(3^x)^2 - 3^2.(3^x) + 18 = 0$ ← $3^{2x} = (3^x)^2$; $3^{x+2} = 3^x.3^2$.

Put $y = 3^x$

$y^2 - 9y + 18 = 0$

$(y - 3)(y - 6) = 0$ ← Factorise.

$y = 3$ or $y = 6$

$3^x = 3$ or $3^x = 6$ ← Resubstitute for y.

$x = 1 \quad \log_{10} 3^x = \log_{10} 6$

$x \log_{10} 3 = \log_{10} 6$

$$x = \frac{\log_{10} 6}{\log_{10} 3} = 1.631$$

(2) (a) $e^{x-2} = 8$ ← Take logs base e on both sides.

$\ln e^{x-2} = \ln 8$

$(x - 2) \ln e = \ln 8$

$x - 2 = \ln 8$ ← $\ln e = 1$.

$x = \ln 8 + 2 = 4.08$ (3 s.f.)

(b) $\log_x 15 = 3$ ← Change to an equation without the logarithm.

$\Rightarrow \quad x^3 = 15$

$\Rightarrow \quad x = \sqrt[3]{15} = 2.47$ (3 s.f.)

(c) $2 \ln (2x - 1) = 5$

$\ln (2x - 1) = 2.5$ ← Isolate the logarithm term.

$\Rightarrow \quad 2x - 1 = e^{2.5}$ ← Raise both sides to the power e.

$\Rightarrow \quad 2x = e^{2.5} + 1$

$\Rightarrow \quad x = 6.59$ (3 s.f.)

(3) $3^x = 5^{2x-3}$ ← To solve an equation where the unknown is a power we usually take logs of both sides.

$\log_{10} 3^x = \log_{10} 5^{2x-3}$

$x\log_{10} 3 = (2x - 3)\log_{10} 5$

$3\log_{10} 5 = 2x\log_{10} 5 - x\log_{10} 3$ ← Collecting the x terms on one side.

$3\log_{10} 5 = x(2\log_{10} 5 - \log_{10} 3)$ ← Factorise.

$$\frac{3\log_{10} 5}{(2\log_{10} 5 - \log_{10} 3)} = x$$

$x = 2.28$ (3 s.f.)

(4) $\sum_{r=1}^{20} \ln [5(3^r)]$

$= \ln (5 \times 3) + \ln (5 \times 3^2) + \ldots + \ln (5 \times 3^{20})$ ← Write out the full series first.

$= (\ln 5 + \ln 3) + (\ln 5 + 2 \ln 3) + \ldots + (\ln 5 + 20 \ln 3)$ ← Using the log laws. This is an arithmetic series, with common difference ln 3.

$$S_{20} = \frac{(\ln 5 + \ln 3) + (\ln 5 + 20 \ln 3)}{2} \times 20$$ ← Using $S = \left(\frac{a + l}{2}\right)n$.

$= 10(2 \ln 5 + 21 \ln 3)$

$= 210 \ln 3 + 20 \ln 5$

Core 2

(5) (a) $\log_{10}(x + 1) - \log_{10}(x - 1) = 1$

$\Rightarrow \quad \log_{10}\left(\dfrac{x+1}{x-1}\right) = 1$ ← Combining the logs.

$\Rightarrow \quad \left(\dfrac{x+1}{x-1}\right) = 10$ ← Raising both sides to the power 10.

$\Rightarrow \quad x + 1 = 10x - 10$

$\Rightarrow \quad 11 = 9x$ ← Collect the x-terms.

$\Rightarrow \quad \frac{11}{9} = x$

$\Rightarrow \quad x = 1.22$ (2 d.p.)

(b) $2 + \ln x = \ln(x + 3)$

$2 = \ln(x + 3) - \ln x$ ← Collect the log terms.

$2 = \ln\left(\dfrac{x+3}{x}\right)$ ← Combine the log terms.

$e^2 = \dfrac{x+3}{x}$ ← Raise both sides to the power 10.

$xe^2 = x + 3$

$xe^2 - x = 3$ ← Collect the x terms.

$(e^2 - 1)x = 3$ ← Factorise.

$x = \dfrac{3}{(e^2 - 1)} = 0.47$ (2 d.p.)

(6) (a) $a = 10 - e^{2\ln 2}$ ← Substituting the coordinates of A into the equation.

$a = 10 - e^{\ln 2^2}$ ← $e^{\ln x} = x$ (and $\ln e^x = x$).

$= 10 - 2^2$

$a = 6$

$-6 = 10 - e^{2b}$ ← Substituting the coordinates of B into the equation.

$e^{2b} = 16$

$e^b = 4$ ← Square rooting both sides.

$\ln e^b = \ln 4$

$b = \ln 4 = 1.39$ (3 s.f.)

(b) Mid-point is $\left(\dfrac{\ln 2 + \ln 4}{2}, \dfrac{6 + -6}{2}\right)$ ← Using $\left(\dfrac{x_1 + x_2}{2}, \dfrac{y_1 + y_2}{2}\right)$

i.e. $\left(\dfrac{\ln 8}{2}, 0\right)$ ← $\ln a + \ln b = \ln ab$

i.e. (1.04, 0) (3 s.f.)

Core 2

2.6 Differentiation

(1) $\frac{dy}{dx} = 3x^2 - 18x + 24$

For turning points, $3x^2 - 18x + 24 = 0$ ← **The gradient is zero at a turning point.**

$x^2 - 6x + 8 = 0$ ← **Simplify before solving.**

$(x - 4)(x - 2) = 0$

$x = 4$ or 2

$y = 16$ or 20 ← **We need both coordinates.**

$\frac{d^2y}{dx^2} = 6x - 18$

When $x = 4$, $\frac{d^2y}{dx^2} = 6$; when $x = 2$, $\frac{d^2y}{dx^2} = -6$ ← **To prove that they are turning points we must show that $\frac{d^2y}{dx^2} \neq 0$ at each point.**

The turning points are (4, 16) (minimum) and (2, 20) (maximum).

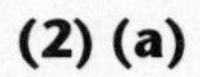

(2) (a)

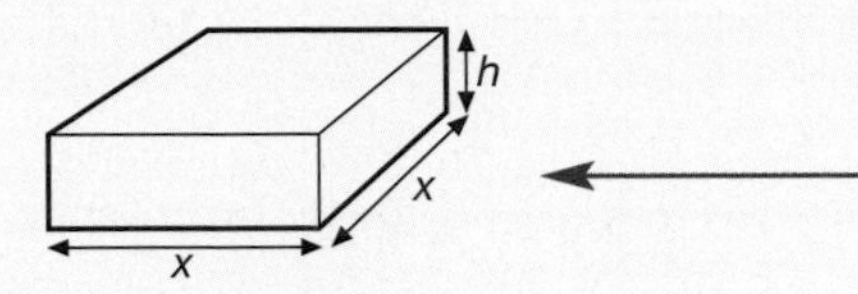

← **It is always a good idea to draw a diagram.**

Let h cm be the height of the tank. ← **A third variable will always be needed; this may sometimes be defined in the question itself.**

Then the volume $hx^2 = 500$ litres $= 500\,000$ cm^3. ← **We need the volume in cm^3 as h and x are in cm.**

Thus $h = \frac{500\,000}{x^2}$ ← **The constraint (here the volume being 500 litres) is used to obtain a connection between the two variables.**

$A = x^2 + 4hx$ ← **The base plus the four sides.**

$= x^2 + 4x\left(\frac{500\,000}{x^2}\right)$ ← **Substituting for h.**

$= x^2 + \frac{2\,000\,000}{x}$ ← **Cancelling the x.**

(b) $A = x^2 + 2\,000\,000x^{-1}$ ← **All x terms need to be written as powers before differentiating.**

$\frac{dA}{dx} = 2x - 2\,000\,000x^{-2}$ ← **Note that $-1 - 1 = -2$!**

Hence, $2x - 2\,000\,000x^{-2} = 0$

$x^3 - 1\,000\,000 = 0$ ← **Dividing by 2 and multiplying through by x^{-2}, to clear the fractions.**

i.e. $x = 100$

so $A = 100^2 + 2\,000\,000\,(100^{-1})$

$= 30\,000$ is the minimum value of A.

(c) $\frac{d^2A}{dx^2} = 2 + 4\,000\,000x^{-3}$

When $x = 100$, $\frac{d^2A}{dx^2} = 6 > 0$, hence it is a minimum.

(3) (a) $x + \frac{1}{4x} = -\frac{5}{4}$

$4x^2 + 1 = -5x$ ← **Multiplying through by $4x$ to clear the fractions.**

$4x^2 + 5x + 1 = 0$

$(4x + 1)(x + 1) = 0$

$x = -\frac{1}{4}$ or -1.

(b) $f(x) = x + \left(\frac{1}{4}\right)x^{-1}$

$f'(x) = 1 - \left(\frac{1}{4}\right)x^{-2}$

$1 - \left(\frac{1}{4}\right)x^{-2} > 0$

$4x^2 - 1 > 0$

$(2x + 1)(2x - 1) > 0$

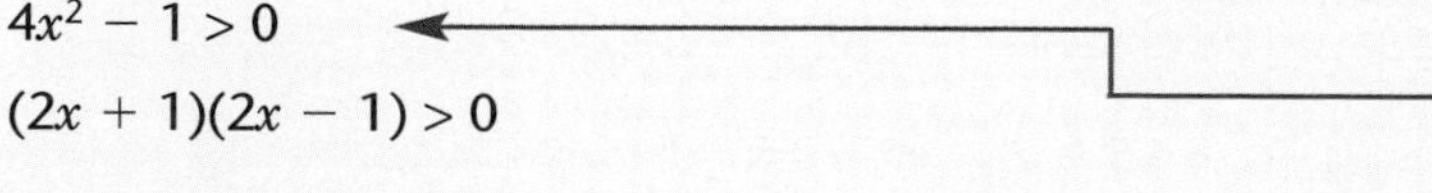

> **Before differentiating, write all terms as powers of x.**
>
> **f is increasing if its gradient is positive.**
>
> **Multiplying through by $4x^2$ to clear the fractions.**

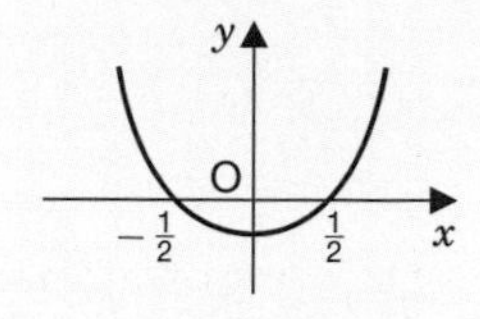

> **To solve a quadratic inequality sketch its graph.**

$x < -\frac{1}{2}$ or $x > \frac{1}{2}$

(4) (a) $y = x^3 + x^2 - x + 1$

$\frac{dy}{dx} = 3x^2 + 2x - 1 = 0$

$(3x - 1)(x + 1) = 0$

$x = \frac{1}{3}$ or $x = -1$

$y = \frac{1}{27} + \frac{1}{9} - \frac{1}{3} + 1$ or $y = -1 + 1 + 1 + 1$

$= \frac{22}{27}$ $\qquad = 2$

$\left(\frac{1}{3}, \frac{22}{27}\right)$; $(-1, 2)$

> **$\frac{dy}{dx} = 0$ at stationary points.**
>
> **Factorise.**
>
> **Solving for x.**
>
> **Substitute to find the y-values.**

(b) $\frac{d^2y}{dx^2} = 6x + 2$

When $x = \frac{1}{3}$, $\frac{d^2y}{dx^2} = 2 + 2 > 0$ i.e. minimum

When $x = -1$, $\frac{d^2y}{dx^2} = -6 + 2 < 0$ i.e. maximum

(c) $\frac{dy}{dx} > 0 \Rightarrow (3x - 1)(x + 1) > 0$

$\Rightarrow x > \frac{1}{3}$ or $x < -1$

> **For an increasing function, gradient > 0.**

(d) When $x = 0$, $y = 1$

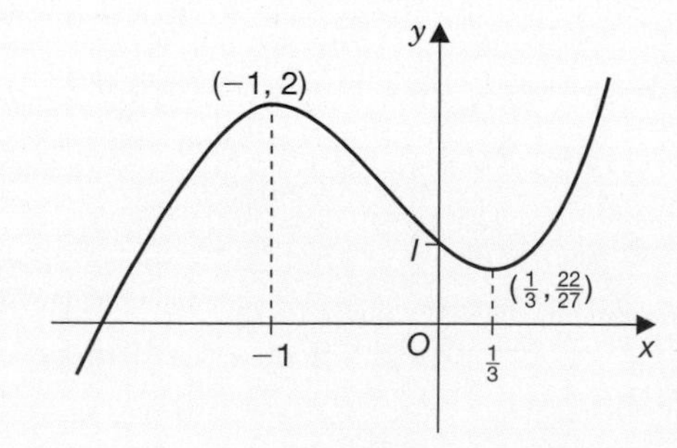

(5) (a) $0 = 4x + 15 + \frac{9}{x}$

$0 = 4x^2 + 15x + 9$

$0 = (4x + 3)(x + 3)$

$\Rightarrow x = -\frac{3}{4}$ or -3

$\Rightarrow \left(-\frac{3}{4}, 0\right)$; $(-3, 0)$

> **Put $y = 0$.**
>
> **Multiply by x to clear fraction.**
>
> **Factorise.**

(b) $f'(x) = 4 - \dfrac{9}{x^2} = 0$ ← **Gradient is 0 at a turning point.**

$4x^2 = 9$ ← **Multiplying by x^2 to clear fraction.**

$x^2 = \frac{9}{4}$

$x = \pm\frac{3}{2}$ ← **Two square roots.**

$y = 6 + 15 + 6$ or $y = -6 + 15 - 6$ ← **Substituting to find x.**

$= 27$ $\quad\quad = 3$

Hence points are $(\frac{3}{2}, 27)$; $(-\frac{3}{2}, 3)$

(6) (a) Let w m be width of enclosure

then $2w + x = 200$ ← **This is the constraint.**

$w = \frac{1}{2}(200 - x)$ ← **We will need to substitute for w.**

$A = xw = \frac{1}{2}(200x - x^2)$

(b) $\dfrac{dA}{dx} = \frac{1}{2}(200 - 2x) = 0$ ← **To maximise find the turning point.**

$\Rightarrow \quad x = 100$

$\dfrac{d^2A}{dx^2} = -1 < 0 \Rightarrow$ maximum

$A_{max} = \frac{1}{2}(200 \times 100 - 100^2)$ ← **Substitute for x.**

$= \dfrac{100}{2}(200 - 100)$

$= 5000$

$\therefore$ Maximum area is 5000 m^2.

(7) $f(x) = 2x - 3 + 8x^{-1}$ ← **Write all terms as powers of x.**

$f'(x) = 2 - 8x^{-2}$

so, $2 - \dfrac{8}{x^2} < 0$ ← **$f'(x) < 0$ for a decreasing function.**

$2x^2 < 8$ ← **Multiply by x^2 to clear fraction. (Note that $x^2 \geqslant 0$ for all x.)**

$x^2 < 4$

$-2 \leqslant x < 2$ ← **Care when square rooting inequalities.**

(8) $f(x) = 3x^{\frac{5}{2}} - 5x^{\frac{3}{2}}$

(a) $f'(x) = \dfrac{15x^{\frac{3}{2}}}{2} - \dfrac{15x^{\frac{1}{2}}}{2}$

$= \dfrac{15x^{\frac{1}{2}}}{2}(x - 1)$

(b) $0 = \dfrac{15x^{\frac{1}{2}}}{2}(x - 1)$ ← **$f'(x) = 0$ at a turning point.**

$\Rightarrow \quad x = 1$ as $x \neq 0$ (given)

$\Rightarrow \quad y = 3 - 5 = -2$ ← **Substitute back to find y-coordinate.**

so, $(1, -2)$ is the turning point

(c) $f''(x) = \dfrac{45x^{\frac{1}{2}}}{4} - \dfrac{15x^{-\frac{1}{2}}}{4}$

$f''(1) = \dfrac{45}{4} - \dfrac{15}{4} = \dfrac{30}{4} > 0$

so, $(1, -2)$ is a minimum point.

2.7 Integration

(1) (a) $\frac{dy}{dx} = 13.5x^{-0.5} - 1.5x^{0.5}$

$= 1.5x^{-0.5}(9 - x)$

(b) $1.5x^{-0.5}(9 - x) = 0 \rightarrow x = 9$ ← **At a turning point the gradient is zero.**

When $x = 9$, $y = 27.(9)^{0.5} - 9^{1.5}$ ← **Both coordinates are required.**

$= 81 - 27 = 54$ i.e. A is (9, 54).

(c) Area $= \int_0^{27} (27x^{0.5} - x^{1.5})\, dx$ ← **Area =** $\int_a^b y\,dx$.

$= [18x^{1.5} - 0.4x^{2.5}]_0^{27}$

$= 1010$ (3 s.f.)

(2) (a) P is (2, 0); Q is (4, 0). ← **'Write down' means marks for answers only.**

(b) $(x - 2) = (x - 2)(x - 4)$ ← **Solving simultaneously.**

$0 = (x - 2)(x - 4 - 1)$ ← **$(x - 2)$ is clearly a common factor.**

$0 = (x - 2)(x - 5)$

$x = 2$ or $x = 5$

When $x = 5$, $y = 3$ i.e. R is (5, 3). ← **P is (2, 0).**

(c) Area $= \int_2^5 ((x - 2) - (x - 2)(x - 4))dx$ ← **Area enclosed =** $\int_a^b y_{top} - y_{bottom}\, dx$

$= \int_2^5 ((x - 2)(5 - x))dx$ ← **Simplify before integrating.**

$= \int_2^5 (-x^2 + 7x - 10)dx$ ← **Multiply out first.**

$= [-\frac{1}{3}x^3 + \frac{7}{2}x^2 - 10x]_2^5$

$= 4.5$

(3) (a) $\frac{dy}{dx} = -2x$ ← **Gradient of the tangent is the gradient of the curve.**

When $x = 1$, $\frac{dy}{dx} = -2$

equation of tangent is $y - 3 = -2(x - 1)$ ← **Using $y - y_1 = m(x - x_1)$.**

i.e. $y = -2x + 5$

$0 = -2x + 5 \rightarrow x = 2.5$ ← **The x-axis is the line $y = 0$.**

(b) The curve crosses the x-axis when $0 = 4 - x^2$

i.e. $x = 2$ (or -2)

Shaded area = Area under tangent between $x = 1$ and $x = 2.5$ − Area under curve between $x = 1$ and $x = 2$

$= \frac{1}{2} \times 1.5 \times 3 - \int_1^2 (4 - x^2)dx$

$= \frac{9}{4} - [4x - \frac{1}{3}x^3]_1^2$

$= \frac{9}{4} - \frac{5}{3}$

$= \frac{7}{12}$

Core 2

(4) (a) $y^2 = (x^{\frac{3}{2}} - 3)^2$

$= x^3 - 6x^{\frac{3}{2}} + 9$

Multiply out brackets.

Don't forget 'middle' term.

(b) $\int_1^4 y^2 dx = \int_1^4 (x^3 - 6x^{\frac{3}{2}} + 9)dx$

$= \left[\frac{1}{4}x^4 - \frac{12x^{\frac{5}{2}}}{5} + 9x\right]_1^4$

Use top-heavy fractions.

$= \frac{1}{4}(4^4 - 1^4) - \frac{12}{5}(4^{\frac{5}{2}} - 1^{\frac{5}{2}}) + 9(4 - 1)$

Evaluating term-by-term.

$= \frac{255}{4} - \frac{12}{5} \times 31 + 27$

$= 16\frac{7}{20}$

(5) (a)

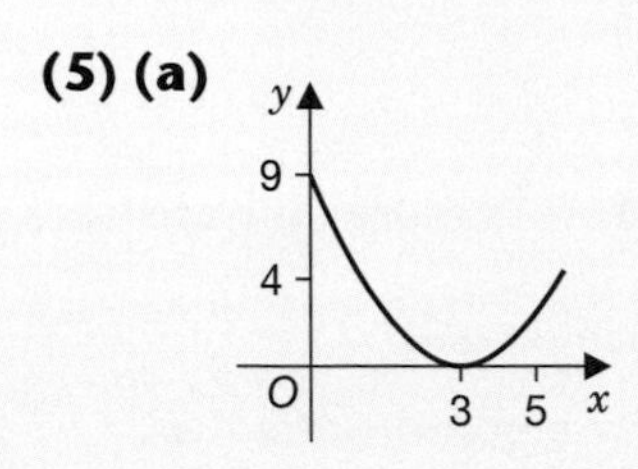

When $y = 0, (x - 3)^2 = 0$

$\Rightarrow \quad x = 3$ (twice)

When $x = 0, y = 9$

$x = 5, y = 4$

The repeated root indicates graph touches the x-axis at $x = 3$.

(b) Area $= \int_0^1 (x - 3)^2 dx$

$= \int_0^1 (x^2 - 6x + 9)dx$

Multiply out first before integrating.

$= \left[\frac{x^3}{3} - 3x^2 + 9x\right]_0^1$

$= \frac{1}{3} - 3 + 9$

$= 6\frac{1}{3}$ square units

(6) $\int_1^k \frac{1}{x^2}dx = \int_k^3 \frac{1}{x^2}dx$

Remember $x^{-2} = \frac{1}{x^2}$

$\Rightarrow \quad \left[\frac{1}{x}\right]_k^1 = \left[\frac{1}{x}\right]_3^k$

$\int x^{-2}dx = -x^{-1} = -\frac{1}{x}$

Change sign and swap limits.

$\Rightarrow \quad 1 - \frac{1}{k} = \frac{1}{k} - \frac{1}{3}$

$\Rightarrow \quad \frac{4}{3} = \frac{2}{k}$

$\Rightarrow \quad 4k = 6$

$\Rightarrow \quad k = \frac{3}{2}$

(7) (a) $0 = 3x^{\frac{5}{2}} - 5x^{\frac{3}{2}}$

$0 = x^{\frac{3}{2}}(3x - 5)$

Factorising.

$\Rightarrow x = \frac{5}{3}$ as $x \neq 0$

(b) $\int_1^4 y\,dx = \int_1^4 (3x^{\frac{5}{2}} - 5x^{\frac{3}{2}})dx$

$= [\frac{6}{7}x^{\frac{7}{2}} - 2x^{\frac{5}{2}}]_1^4$

Use top-heavy fractions when integrating.

$= \frac{6}{7}(4^{\frac{7}{2}} - 1^{\frac{7}{2}}) - 2(4^{\frac{5}{2}} - 1^{\frac{5}{2}})$

Evaluating term-by-term.

$= \frac{6}{7} \times (128 - 1) - 2(32 - 1)$

$= (\frac{6}{7} \times 127) - 62$

$= \dfrac{762 - 434}{7}$

Taking common denominator.

$= \dfrac{328}{7} = 46\frac{6}{7}$

(8) (a) $p = x^{\frac{1}{2}} \Rightarrow p^2 = x$

so, $p^2 - 3p + 2 = 0$

$\Rightarrow (p - 2)(p - 1) = 0$

Factorising.

$\Rightarrow p = 2$ or $p = 1$

i.e. $x^{\frac{1}{2}} = 2$ or $x^{\frac{1}{2}} = 1$

Substitute back.

$\Rightarrow x = 4$ or $x = 1$

Squaring to give x.

(b) (1, 0) and (4, 0)

Using the answers to part (a).

(c) Area $= \int_1^4 y\,dx$

$= \int_1^4 (x - 3x^{\frac{1}{2}} + 2)dx$

$= \left[\dfrac{x^2}{2} - 2x^{\frac{3}{2}} + 2x\right]_1^4$

Using the standard form.

$= \frac{1}{2}(4^2 - 1^2) - 2(4^{\frac{3}{2}} - 1^{\frac{3}{2}}) + 2(4 - 1)$

Evaluating term-by-term.

$= \dfrac{15}{2} - \dfrac{14}{3} + 6$

$= 8\frac{5}{6}$

Core 2

(9) (a) $f(x) = x(x^2 - 2)(3x - x^{-1})$

$= (x^2 - 2)(3x^2 - 1)$

Multiplying the first and third terms is easier.

$= 3x^4 - 7x^2 + 2$

(b) $\int_0^2 f(x)dx = \int_0^2 (3x^4 - 7x^2 + 2)dx$

Using the multiplied out version.

$= \left[\dfrac{3x^5}{5} - \dfrac{7x^3}{3} + 2x\right]_0^2$

Integrating term-by-term.

$= \dfrac{96}{5} - \dfrac{56}{3} + 4$

$= 4\frac{8}{15}$

Chapter 3

Mechanics 1

Questions with model answers

C grade candidate – mark scored 9/15

(1) At 12 noon ship S has position vector $(-9\mathbf{i} + 6\mathbf{j})$ km and is moving with constant velocity $(3\mathbf{i} + 12\mathbf{j})$ km h^{-1} and ship T has position vector $(16\mathbf{i} + 6\mathbf{j})$ km and is moving with constant velocity $(-9\mathbf{i} + 3\mathbf{j})$ km h^{-1}.

(a) Find how far apart the ships are at 12 noon. [4]

$16\mathbf{i} + 6\mathbf{j} - -9\mathbf{i} + 6\mathbf{j} = 25\mathbf{i} + 12\mathbf{j}$

Distance $= \sqrt{(25^2 + 12^2)} = 27.7$ Km

(b) Write down the position vectors of S and T at time t hours after noon and hence find the vector from S to T at time t hours after noon. [5]

$\mathbf{s} = (-9\mathbf{i} + 6\mathbf{j}) + t(3\mathbf{i} + 12\mathbf{j})$

$\mathbf{t} = (16\mathbf{i} + 6\mathbf{j}) + t(-9\mathbf{i} + 3\mathbf{j})$

$\overrightarrow{ST} = \mathbf{t} - \mathbf{s} = (16\mathbf{i} + 6\mathbf{j}) + t(-9\mathbf{i} + 3\mathbf{j}) - \{(-9\mathbf{i} + 6\mathbf{j}) + t(3\mathbf{i} + 12\mathbf{j})\}$

$= 25\mathbf{i} + t(-12\mathbf{i} - 9\mathbf{j})$

(c) Find the least distance between the ships in the subsequent motion. [6]

$\overrightarrow{ST} = 25\mathbf{i} + t(-12\mathbf{i} - 9\mathbf{j}) = (25 - 12t)\mathbf{i} - 9t\mathbf{j}$

$|\overrightarrow{ST}| = (25 - 12t)^2 + 81t^2$

$= 625 + 144t^2 + 81t^2$

$y = 625 + 225t^2$

$\frac{dy}{dx} = 450t = 0$

$t = 0$

They are closest together at the start.

Examiner's Commentary

(a) Incorrect – the brackets around the second vector have been omitted resulting in the **j** terms being added instead of being subtracted, **2/4 scored.**

(b) Correct

Correct

The brackets this time are remembered! Putting t = 0 in the answer would have revealed the error in part (a), **5/5 scored.**

(c) A good start – collecting the i's and j's.

Also correct.

Incorrect algebra – 600t has been missed out.

Wrong notation – not dy/dx but dy/dt – but the method is correct.

Correct conclusion from wrong working, **2/6 scored.**

?

For help see Revise AS Study Guide section 3

GRADE BOOSTER

When subtracting vectors in component form ensure that the second vector goes in brackets.
e.g. 16i + 6j – (–9i + 6j)

A grade candidate – mark scored 11/14

Examiner's Commentary

(1) A particle P of mass 2 kg is acted upon by two horizontal forces $(2\mathbf{i} + 3\mathbf{j})$N and $(4\mathbf{i} - 7\mathbf{j})$N where **i** and **j** are unit horizontal vectors due East and due North respectively.
Find

(a) the magnitude of the acceleration of P: [4]

$(2\mathbf{i} + 3\mathbf{j}) + (4\mathbf{i} - 7\mathbf{j}) = 2\mathbf{a}$ — Applying $F = ma$.

$3\mathbf{i} - 2\mathbf{j} = \mathbf{a}$

$|\mathbf{a}| = \sqrt{3^2 + (-2)^2} = \sqrt{13} = 3.61$ — Using Pythagoras.

Magnitude of acceleration is $3.61\,\text{m s}^{-2}$ — All correct, **4/4 scored.**

(b) the direction of the acceleration of P: [3]

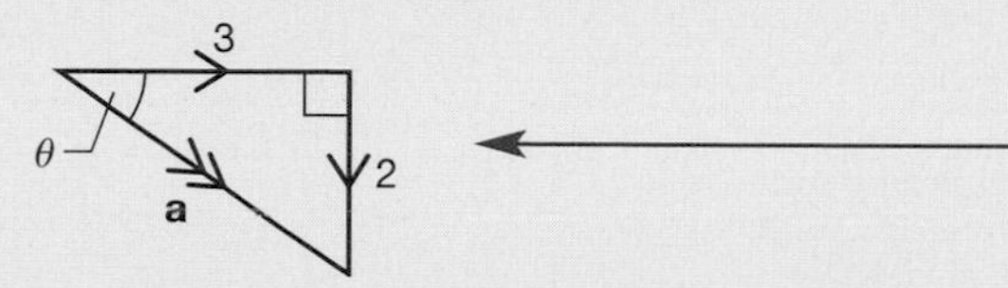

A diagram is a good idea.

$\tan\theta = \frac{2}{3}$

$\theta = 33.7°$. The direction of acceleration is 33.7° S of East. — All correct, **3/3 scored.**

At time $t = 0$, P is at the point with position vector $(\mathbf{i} - \mathbf{j})$ m and is moving with velocity $(\mathbf{i} + \mathbf{j})\,\text{m s}^{-1}$.
Find, when $t = 4$ s,

(c) the speed of P: [4]

$\mathbf{v} = (\mathbf{i} + \mathbf{j}) + 4(3\mathbf{i} - 2\mathbf{j}) = 13\mathbf{i} - 7\mathbf{j}$ — Using $\mathbf{v} = \mathbf{u} + \mathbf{a}t$, but the candidate forgets to find the speed, i.e. $|\mathbf{v}|$, **2/4 scored.**

(d) the position vector of P: [3]

$\mathbf{r} = 4(\mathbf{i} + \mathbf{j}) + \frac{1}{2}(3\mathbf{i} - 2\mathbf{j})4^2$ — Using $\mathbf{s} = \mathbf{u}t + \frac{1}{2}\mathbf{a}t^2$.

$= 4\mathbf{i} + 4\mathbf{j} + 24\mathbf{i} - 16\mathbf{j}$

$= 28\mathbf{i} - 12\mathbf{j}$ — Correct so far but the candidate forgets to add his answer onto the initial position vector $(\mathbf{i} - \mathbf{j})$, **2/3 scored.**

For help see Revise AS Study Guide section 3

Exam practice questions

3.1 Vectors

1 Three forces $\mathbf{F}_1$, $\mathbf{F}_2$ and $\mathbf{F}_3$ act on a particle.

$\mathbf{F}_1 = (2\mathbf{i} + 3a\mathbf{j})\text{N}$, $\mathbf{F}_2 = (a\mathbf{i} + b\mathbf{j})\text{N}$, $\mathbf{F}_3 = (b\mathbf{i} + 4\mathbf{j})\text{N}$

Given that the particle is in equilibrium, find the values of a and b. [7]

2 Two forces, of magnitude P N and Q N, have a resultant of magnitude $2\sqrt{5}$ N when the angle between their lines of action is 90°. When the angle between their lines of action is 60° the magnitude of their resultant is $2\sqrt{7}$ N. Find the values of P and Q. [12]

3 A particle is acted upon by two forces $\mathbf{F}_1$ and $\mathbf{F}_2$.

$\mathbf{F}_1 = (2\mathbf{i} + \mathbf{j})\text{N}$ and $\mathbf{F}_2 = (a\mathbf{i} + 2a\mathbf{j})\text{N}$.

(a) Find the angle between $\mathbf{F}_1$ and $\mathbf{j}$. [2]

The resultant $\mathbf{R}$ of $\mathbf{F}_1$ and $\mathbf{F}_2$ is parallel to $\mathbf{i}$.

(b) Find the magnitude of $\mathbf{R}$. [6]

3.2 Kinematics

(Take $g = 9.8 \text{ m s}^{-2}$)

1 A cricket ball is thrown vertically upwards from ground level and takes 5 s to reach the ground again. Find the maximum height of the ball above the ground. [6]

2 A car moves with constant acceleration from a speed of 14 m s^{-1} to a speed of 34 m s^{-1} in 20 s.

(a) Find how far the car travels during this 20 s period. [2]

(b) Find how long it takes to cover half of this distance. [7]

3 A small stone is dropped from the top of a tower. One second later another small stone is thrown vertically downwards from the same point at 19.6 m s^{-1}. Given that the two stones hit the ground at the same time, and assuming no air resistance, calculate the height of the tower. [7]

4 A car travelling along a straight road takes two minutes to travel between two sets of traffic lights which are 2145 m apart. It starts at rest and accelerates uniformly for 30 s. It then moves with constant speed before uniformly decelerating to rest for the final 15 s.

(a) Illustrate the motion on a velocity–time graph. [2]

(b) Find the acceleration of the car. [5]

Answers on pages 64–69 Answers on pages 64–69 Answers on pages 64–69

5 Two cars A and B move along a straight horizontal road with constant acceleration. Car A has acceleration 2 m s^{-2} and B has acceleration 1 m s^{-2}.
At time $t = 0$ car A has speed 1 m s^{-1} and is at the point O, and at $t = 4$ s car B is at O and has speed 16 m s^{-1}. Find

(a) the times between which car B is ahead of car A, [10]

(b) the distance from O at which car B overtakes car A, [2]

(c) the distance from O at which car A overtakes car B. [2]

6 A stone is projected with speed u m s^{-1} at an angle α above the horizontal from the top of a vertical cliff which is 56 m high. The stone moves in a vertical plane which is perpendicular to the cliff and lands in the sea 4 s later at a point which is 32 m from the foot of the cliff. Find

(a) the value of u, [4]

(b) the value of α, [4]

(c) the speed with which the stone hits the sea, [5]

(d) the direction of motion of the stone when it hits the sea. [3]

7 A ball is projected with speed 49 m s^{-1} at an angle of 30° above the horizontal from the top of a vertical cliff which is 196 m high. The ball moves in a vertical plane which is perpendicular to the cliff and lands in the sea at the point P. Find

(a) the greatest height of the ball above the sea, [4]

(b) the time taken to hit the sea, [7]

(c) the distance of P from the foot of the cliff. [2]

8 A train travels in a straight line between two railway stations A and B, stopping at both. There is a signal box S between the two stations and the train passes S at exactly 12 noon. The velocity of the train, v km h^{-1}, at t minutes past noon is given by

$$v = \tfrac{15}{2}(3 + 4t - 4t^2).$$

Find

(a) the velocity of the train as it passes the signal box, [1]

(b) the time of departure from A and arrival at B, [6]

(c) the maximum velocity of the train, [5]

(d) the average velocity of the train between A and B. [6]

Answers on pages 64–69 Answers on pages 64–69 Answers on pages 64–69

3.3 Statics

(Take $g = 9.8 \text{ m s}^{-2}$)

1 A particle is suspended by two light inextensible strings and hangs in equilibrium. One string is inclined at 60° to the horizontal and the tension in that string is 20 N. The other string is inclined at 30° to the horizontal. Find, to 3 significant figures,

(a) the weight of the particle, [3]

(b) the tension in the second string. [3]

2 A particle is placed on a rough plane inclined at an angle α to the horizontal, where $\tan\alpha = \frac{3}{4}$. The particle is kept in equilibrium by a horizontal force of magnitude 10 N, acting in the vertical plane containing the line of greatest slope of the inclined plane through the particle. The coefficient of friction between the particle and the plane is $\frac{1}{2}$. Given that the particle is on the point of slipping up the plane, find the normal reaction of the plane on the particle, and the weight of the particle. [9]

3 A uniform beam AB of mass 10 kg and length 2.4 m is at rest in equilibrium in a horizontal position. The beam is supported by two vertical ropes attached to the beam at the points X and Y where $AX = 0.4$ m and $YB = 0.6$ m. Find the tension in each rope. [6]

4 A non-uniform rod AB has length $3a$ and mass $5m$. It rests in equilibrium in a horizontal position on two supports at the points P and Q, where $AP = PQ = QB = a$. A particle of mass $2m$ is fixed to the rod at B. Given that the rod is on the point of tilting about Q, find the distance of the centre of mass of the rod from B. [5]

Answers on pages 64–69 **Answers** on pages 64–69 **Answers** on pages 64–69

3.4 Dynamics

(Take $g = 9.8 \text{ m s}^{-2}$)

1

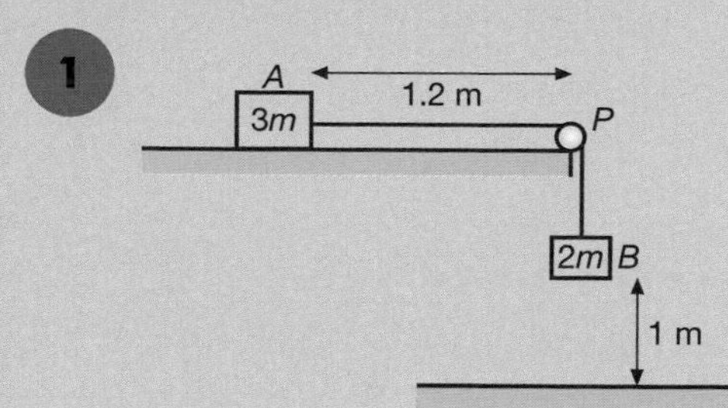

A particle A of mass $3m$ is placed on a rough horizontal table. The particle is attached to one end of a light inextensible string which passes over a small smooth fixed pulley P at the edge of the table. Another particle B of mass $2m$ is attached to the other end of the string and B hangs freely. AP is perpendicular to the edge of the table and A, P and B lie in the same vertical plane. The system is released from rest, with the string taut, when A is 1.2 m from the edge of the table and B is 1 m above the floor, as shown in the diagram.

Given that B strikes the floor after 2 s and does not rebound, find

(a) the acceleration of A during the first two seconds of the motion, [2]

(b) to 2 decimal places, the coefficient of friction between A and the table, [10]

(c) by calculation, whether the particle A reaches the pulley. [6]

2 A particle P of mass $2m$ is moving on a smooth horizontal plane with a speed u when it collides with another particle Q of mass km whose speed is $3u$ in the opposite direction. As a result of the collision the direction of motion of both particles is reversed and the speed of P is halved.

(a) Find the range of values of k. [5]

(b) Find the magnitude of the impulse on P from Q. [3]

3 A particle P of mass 2 kg is pushed, by a constant horizontal force of magnitude 30 N, up a rough plane inclined to the horizontal at an angle α, where $\tan \alpha = \frac{3}{4}$. The particle P moves with constant acceleration $a \text{ m s}^{-2}$. The coefficient of friction between the particle and the inclined plane is 0.2.

(a) Find the magnitude of the normal reaction between the particle and the inclined plane. [4]

(b) Find the value of a. [4]

Answers on pages 64–69 Answers on pages 64–69 Answers on pages 64–69

Answers

3.1 Vectors

examiner's tips

(1) $(2\mathbf{i} + 3a\mathbf{j}) + (a\mathbf{i} + b\mathbf{j}) + (b\mathbf{i} + 4\mathbf{j}) = \mathbf{0}$

> Since the forces are in equilibrium, their resultant is zero.

$(2 + a + b)\,\mathbf{i} + (3a + b + 4)\mathbf{j} = 0\mathbf{i} + 0\mathbf{j}$

> Adding the **i**-components and **j**-components.

$\Rightarrow \quad (2 + a + b) = 0$ and

$(3a + b + 4) = 0$

> Equating coefficients of **i** and **j**.

i.e. $\quad a + b = -2$

$3a + b = -4$

> Subtract the equations to eliminate b.

$\Rightarrow 2a = -2 \Rightarrow a = -1 \Rightarrow b = -1.$

(2)

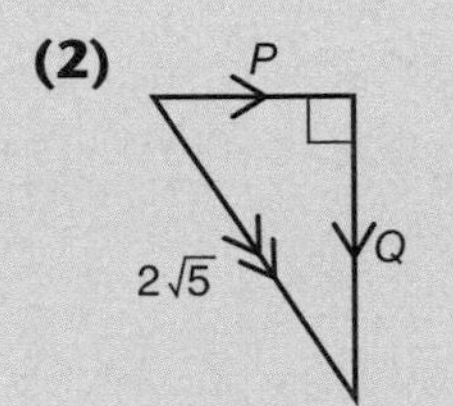

> Draw a vector triangle.

$P^2 + Q^2 = (2\sqrt{5})^2 = 20$

> By Pythagoras.

> Draw another vector triangle, with the arrows 'following each other'; note that the angle is 120° not 60°.

$P^2 + Q^2 - 2PQ\cos 120° = (2\sqrt{7})^2 = 28$

> By cosine rule.

$P^2 + Q^2 + PQ = 28$

> Cos120° = –0.5.

So, $\quad P^2 + Q^2 = 20$ and

$P^2 + Q^2 + PQ = 28$

$\Rightarrow \quad PQ = 8$

> Subtracting the equations.

So, $\quad P^2 + Q^2 + 2PQ = 36$

> Adding – this gives a neat solution as 36 is a perfect square.

$\Rightarrow \quad (P + Q)^2 = 36$

$\Rightarrow \quad P + Q = 6$ i.e. $Q = 6 - P$

> $P + Q = -6$ is impossible as P and Q are magnitudes.

So, $\quad P(6 - P) = 8$

> Substituting for Q in $PQ = 8$.

$\Rightarrow \quad P^2 - 6P + 8 = 0$

$\Rightarrow \quad (P - 2)(P - 4) = 0 \Rightarrow P = 2$ or 4

$Q = 4$ or 2

> There are two symmetrical solutions possible.

(3) (a)

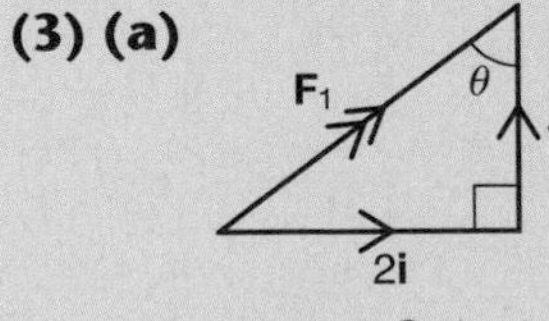

> A simple diagram is all that is needed.

$\tan\theta = 2 \Rightarrow \theta = 63.4°$

(b) $\mathbf{R} = (2\mathbf{i} + \mathbf{j}) + (a\mathbf{i} + 2a\mathbf{j}) = (2 + a)\mathbf{i} + (1 + 2a)\mathbf{j}$

So, $\quad (1 + 2a) = 0$ i.e. $a = -0.5$

> ***R*** parallel to **i** → **j**-component of ***R*** is zero.

So, $\quad \mathbf{R} = 1.5\mathbf{i}$

Hence, $\quad |\mathbf{R}| = 1.5$

The magnitude of **R** is 1.5 N.

3.2 Kinematics

(1) (↑) $0 = 5u - 4.9 \times 5^2 \Rightarrow u = 24.5$

Choose a positive direction, here upwards. Using $s = ut + \frac{1}{2}at^2$.

(↑) $0 = 24.5^2 - 2gh \Rightarrow h = 30.625$

Using $v^2 = u^2 + 2as$.

Maximum height is 30.625 m.

Do not include units in your working but give the final answer with units.

(2) (a) $s = \frac{1}{2}(14 + 34)20 = 480$

Using $s = \frac{1}{2}(u + v)t$.

The car travels 480 m.

(b) $34 = 14 + 20a \Rightarrow a = 1$

First we need to find a, using $v = u + at$.

$240 = 14t + \frac{1}{2}t^2$

Using $s = ut + \frac{1}{2}at^2$.

$t^2 + 28t - 480 = 0$

Multiplying through by 2 to clear fractions and collecting terms.

$(t + 40)(t - 12) = 0$

$t = -40$ or $t = 12$

Include both solutions and then reject where appropriate.

The car takes 12 s.

(3) Suppose the first stone hits the ground after t sec.

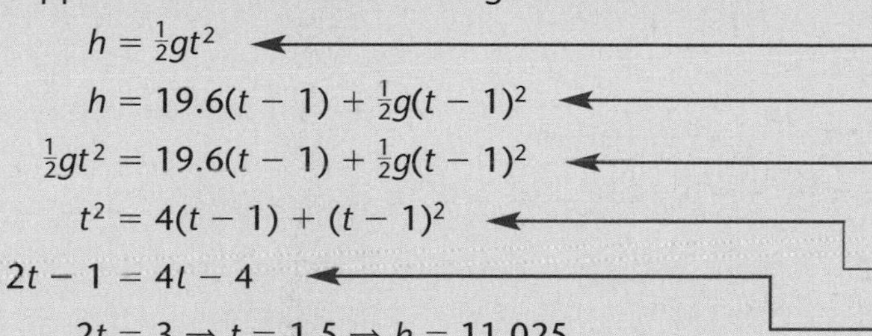

$h = \frac{1}{2}gt^2$

Using $s = ut + \frac{1}{2}at^2$ for the first stone.

$h = 19.6(t - 1) + \frac{1}{2}g(t - 1)^2$

Using $s = ut + \frac{1}{2}at^2$ for the second stone.

$\frac{1}{2}gt^2 = 19.6(t - 1) + \frac{1}{2}g(t - 1)^2$

Although we want h, it is easier to find t and then find h.

$t^2 = 4(t - 1) + (t - 1)^2$

Dividing through by g.

$2t - 1 = 4t - 4$

Multiplying out and cancelling terms.

$2t = 3 \Rightarrow t = 1.5 \Rightarrow h = 11.025$

The height of the tower is 11.025 m.

(4) (a)

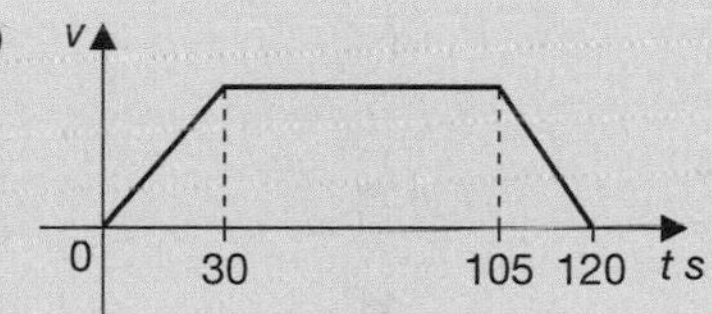

(b) Distance $= 2145 \Rightarrow$ area $= 2145$

$\Rightarrow \frac{1}{2}(120 + 75)v = 2145$

Using the Trapezium Rule.

$\Rightarrow 195v = 4290$

$\Rightarrow v = 22$

So acceleration is $\frac{22}{30} = 0.73 \text{ m s}^{-2}$ (to 2 d.p.).

(5) (a) $s_A = t + \frac{1}{2} \times 2t^2 = t + t^2$

$s_B = 16(t - 4) + \frac{1}{2}(t - 4)^2$

Using $s = ut + \frac{1}{2}at^2$ for both A and B.

We need $s_B > s_A$

Use an inequality rather than an equation.

i.e. $16(t - 4) + \frac{1}{2}(t - 4)^2 > t + t^2$

$32t - 128 + t^2 - 8t + 16 > 2t + 2t^2$

Multiplying through by 2 and expanding the brackets.

$0 > t^2 - 22t + 112$

$0 > (t - 8)(t - 14)$

Use a sketch graph to solve this quadratic inequality.

$8 < t < 14$

i.e. B overtakes A after 8 s and A overtakes B after 14 s.

(b) When $t = 8$, $s_A = 8 + 8^2 = 72$
Cars are 72 m from O.

Using the result above in (a).

(c) When $t = 14$, $s_B = 14 + 14^2 = 210$
Cars are 210 m from O.

Using the result above in (a).

(6) (a) and **(b)**

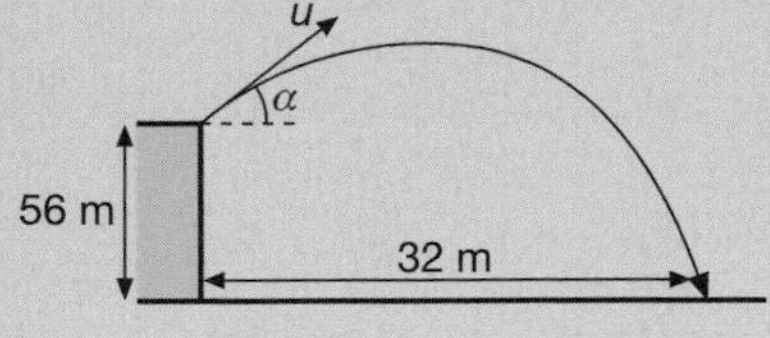

A simple diagram helps to clarify the situation.

$(\rightarrow)\ 32 = 4u\cos\alpha \quad \Rightarrow u\cos\alpha = 8$

Using $s = ut$ ($a = 0$ horizontally).

$(\uparrow) - 56 = 4u\sin\alpha - \frac{1}{2} \times 9.8 \times 4^2$

$\Rightarrow u\sin\alpha = 5.6$

Using $s = ut + \frac{1}{2}at^2$ with upwards positive.

So $\tan\alpha = \dfrac{5.6}{8} \quad \Rightarrow \alpha = 35.0°$

and $u = 9.77$

Dividing the two equations to eliminate u.

(c) $(\uparrow)\ v = u\sin\alpha - 9.8 \times 4$

$= 5.6 - 39.2 = -33.6$

First find the vertical component of the velocity when the stone hits the sea, using $v = u + at$; the negative sign indicates it is going downwards.

So speed $= \sqrt{8^2 + 33.6^2}\ \text{ms}^{-1}$

$= 34.5\ \text{m s}^{-1}$

The horizontal component is 8 from above.

(d)

8
θ
33.6

A simple diagram helps to clarify the situation.

$\tan\theta = \dfrac{33.6}{8} \Rightarrow \theta = 76.6°$ (3 s.f.)

The stone hits the sea at 76.6° to the horizontal.

(7) (a)

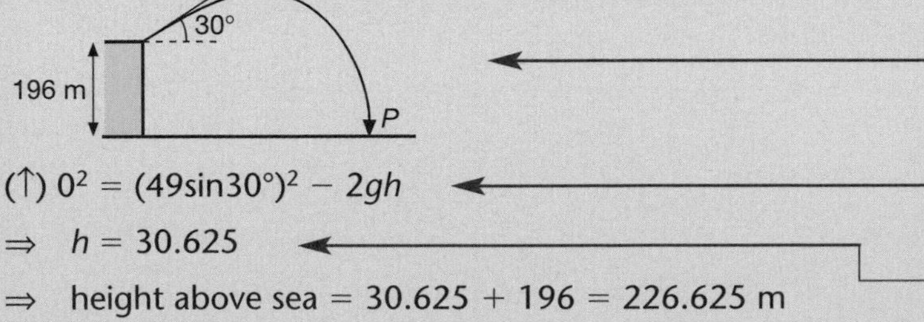

A simple diagram is a good idea.

$(\uparrow)\ 0^2 = (49\sin 30°)^2 - 2gh$

Applying $v^2 = u^2 + 2as$ vertically, upwards positive.

$\Rightarrow \quad h = 30.625$

This is the height above the point of projection.

$\Rightarrow \quad$ height above sea $= 30.625 + 196 = 226.625$ m

(b) $-196 = 49\sin 30° t - \frac{1}{2}gt^2$

Note $s = -196$; s is displacement not distance travelled.

$t^2 - 5t - 40 = 0$

Dividing through by –4.9 and collecting terms.

$t = \frac{1}{2}(5 \pm \sqrt{185})$

It doesn't factorise so use the formula.

i.e. $\quad t = 9.3$ or -4.3

Ball hits sea after 9.3 s.

(c) $s = 49\cos 30° \times 9.3 = 395$
P is 395 m from the foot of the cliff.

Using $s = ut$ ($a = 0$ horizontally).

(8) (a) When $t = 0$, $v = \dfrac{45}{2} = 22.5$ km h^{-1}.

(b) $0 = \frac{15}{2}(3 + 4t - 4t^2)$

The train stops at both so put $v = 0$.

i.e. $4t^2 - 4t - 3 = 0$

$(2t + 1)(2t - 3) = 0$

$t = -0.5$ or 1.5

The train leaves A at 11.59.5 am and arrives at B at 12.01.5 pm.

(c) $a = \dfrac{dv}{dt} = \dfrac{15}{2}(4 - 8t) = 0$ at maximum velocity

$\Rightarrow \quad t = 0.5 \Rightarrow v_{max} = \dfrac{15}{2}(3 + 2 - 1) = 30$ km h^{-1}

(d) $s = \displaystyle\int \left(\frac{15}{2}(3 + 4t - 4t^2)\right) dt = \frac{15}{2}\left(3t + 2t^2 - \frac{4}{3}t^3\right) + c$

We need to find the distance *AB*.

When $t = -0.5$, $s = 0 \Rightarrow c = \dfrac{25}{4}$

When $t = 1.5$, $s = 40 \dfrac{\text{km}}{60} = \dfrac{2}{3}$ km

Note that because the time units are different, s is in $\dfrac{km}{60}$.

Average velocity $= \dfrac{2}{3} \times 30$ km h^{-1} $= 20$ km h^{-1}.

3.3 Statics

(1) (a)

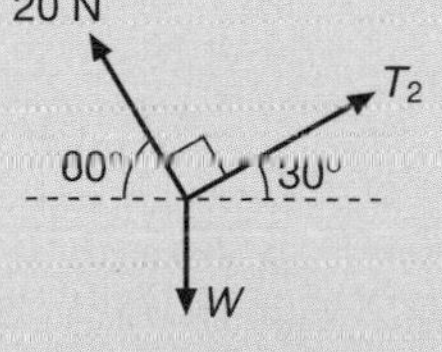

A clear force diagram is essential.

Resolving along the first string,

$20 = W\cos 30° \Rightarrow W = 23.1$ N (3 s.f.).

Exploiting the fact that the two strings are perpendicular – if this were not the case then we would normally resolve vertically and horizontally and then have to solve two simultaneous equations.

(b) Resolving along the second string,

$T_2 = W\cos 60°$

$\Rightarrow T_2 = 23.094 \times 0.5 = 11.5$ N (3 s.f.).

To obtain 3 s.f. answers the working should be to at least 4 s.f.

(2)

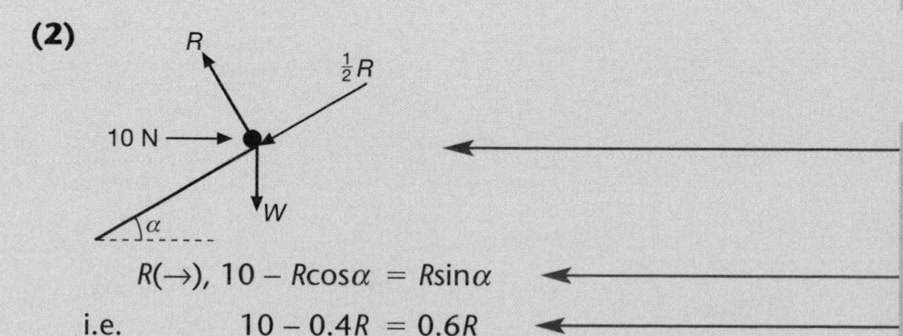

Since the particle is about to slip up the plane, the friction is limiting and is acting down the plane.

$R(\rightarrow)$, $10 - R\cos\alpha = R\sin\alpha$

Resolving horizontally.

i.e. $10 - 0.4R = 0.6R$

$\sin\alpha = 0.6$ and $\cos\alpha = 0.8$.

$R = 10$ N

$R(\uparrow)$, $-\frac{1}{2}R\sin\alpha + R\cos\alpha = W$

Resolving vertically.

$-5 \times 0.6 + 10 \times 0.8 = W$

$5 \text{ N} = W$

(3)

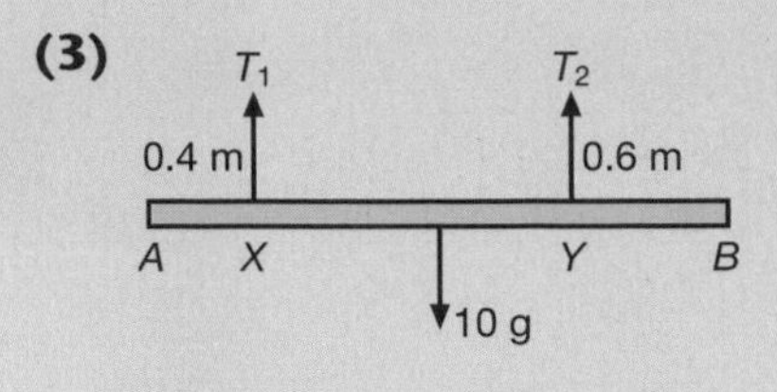

A simple diagram showing all the forces is essential.

$R(\uparrow)$, $T_1 + T_2 = 10g = 98$

Since the rod is in equilibrium we can resolve in any direction.

$M(X)$, $1.4\ T_2 = 10g \times 0.8$

Taking moments about a point through which an unknown force passes is usually a good idea.

$T_2 = 56 \Rightarrow T_1 = 42$

The tensions are 56 N and 42 N.

(4) Let the centre of mass of the rod be x m from Q.

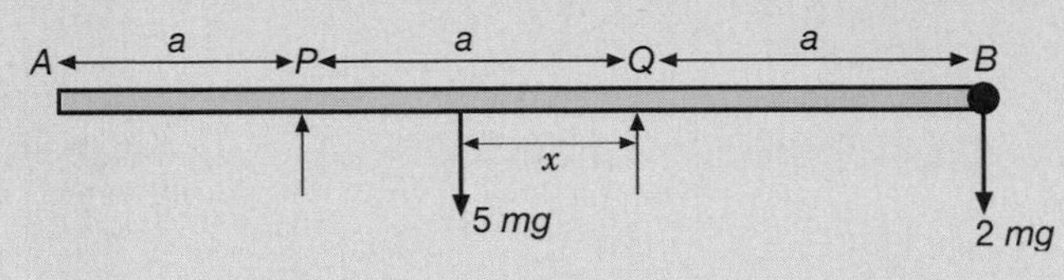

A simple diagram is essential.

$M(Q)$, $5\ mgx = 2\ mga$

$$x = \frac{2a}{5}$$

When the rod is about to tilt about Q the reaction at P will be zero since the contact at P will be about to disappear.

Hence the distance of the centre of mass of the rod from B is $\frac{7a}{5}$.

3.4 Dynamics

(1) (a)

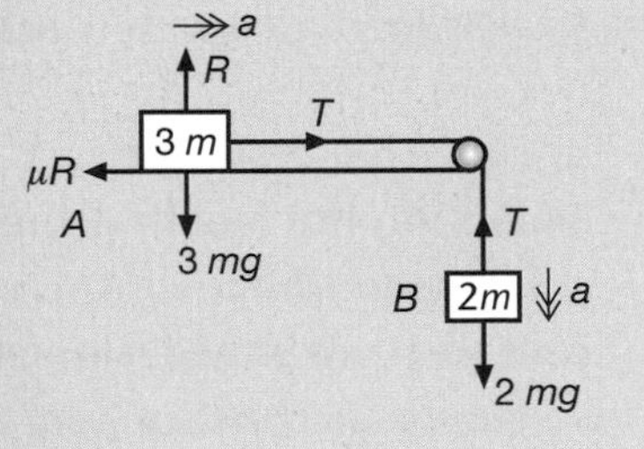

A clear force diagram, showing all the forces and any accelerations is absolutely crucial to the solution.

For B,

$1 = \frac{1}{2}a2^2 \rightarrow a = 0.5$

Using $s = ut + \frac{1}{2}at^2$.

(b) For A,

$$R = 3mg$$

$$T - \mu R = 3ma$$

i.e. $T - \mu 3mg = 3ma$ ......**[1]**

We must consider each mass separately; friction is limiting as there is motion.

For B,

$2mg - T = 2ma$ ......**[2]**

Note that we have resolved in the direction of the acceleration for each mass.

[1] + **[2]**,

$2mg - \mu 3mg = 5ma$

i.e. $g(2 - 3\mu) = 5a$ ......**[3]**

From part **(a)**

$9.8(2 - 3\mu) = 2.5$

$\Rightarrow \quad \mu = 0.58$

(c) B hits the floor with speed $v = 0.5 \times 2 = 1$ ← **Using $v = u + at$.**

For A, $-\mu 3mg = 3ma \Rightarrow a = -\mu g$ ← **A common error is to put 'ma' on the RHS.**

$0 = 1^2 - 2\mu gs \Rightarrow s = 0.088$ ← **Since this is less than 0.2 m A does not reach the pulley.**

(2) (a)

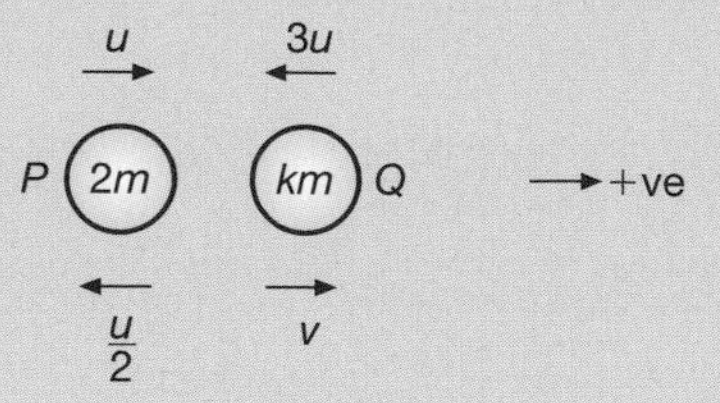

$u \rightarrow$; $3u \leftarrow$

P (2m) (km) Q $\rightarrow$ +ve

$\frac{u}{2} \leftarrow$; $v \rightarrow$

← **A clear diagram showing all the information is essential.**

$$2mu - km3u = -2m\left(\frac{u}{2}\right) + kmv$$ ← **Using conservation of momentum.**

$$3mu - 3kmu = kmv$$

$$3u(1 - k) = kv$$

$$v > 0 \Rightarrow (1 - k) > 0 \Rightarrow k < 1$$ ← **Using the fact that the direction of motion of Q is reversed (i.e. $v > 0$).**

(b)

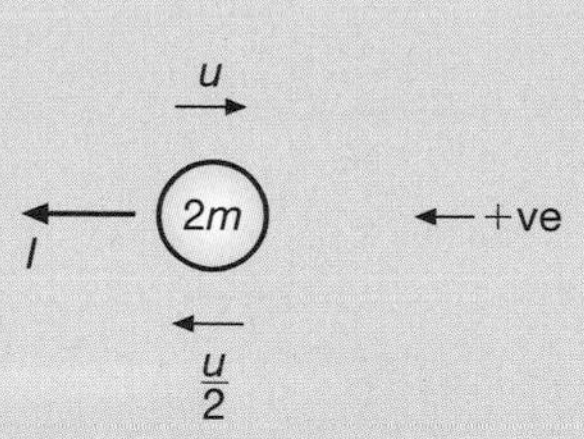

$u \rightarrow$

$I \leftarrow$ (2m) $\leftarrow$ +ve

$\frac{u}{2} \leftarrow$

← **Draw a diagram showing the velocities and the impulse.**

$$I = 2m\frac{u}{2} - (-2mu)$$ ← **Note that the impulse-momentum equation is a vector equation and attention must be paid to signs.**

$$= 3mu$$

(3) (a)

R, a, 30 N, μR, $2g$, α

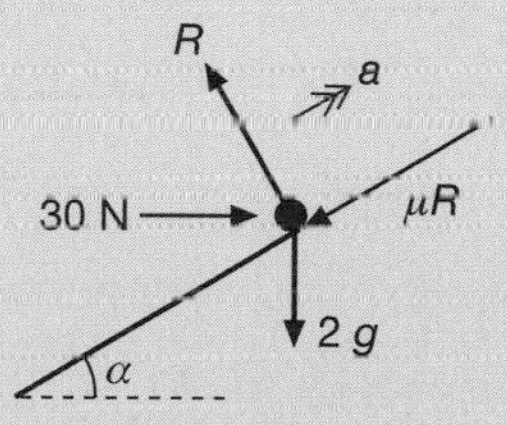

← **Draw a clear force diagram showing the forces and the acceleration.**

$R(\nwarrow)$, $R - 30\sin\alpha - 2g\cos\alpha = 0$ ← **Resolving perpendicular to the acceleration.**

$\Rightarrow R = 18 + 15.68 = 33.68$

Normal reaction is 33.7 N (3 s.f.).

(b) $R(\nearrow)$, $30\cos\alpha - 0.2R - 2g\sin\alpha = 2a$ ← **Resolving parallel to the acceleration.**

$24 - 6.736 - 11.76 = 2a$

$a = 2.75$ (3 s.f.).

Chapter 4

Statistics 1

Questions with model answers

C grade candidate – mark scored 4/7

(1) A consumer group is investigating the use of the telephone within a particular town. The numbers of telephone units, t, used in a particular three-month period by a random sample of 250 households were collated and summarised in a group frequency table. In order to simplify the arithmetic the data in the table were coded such that $x = \frac{1}{10}(t - 290)$, giving

$\sum fx = -40$ and $\sum fx^2 = 1075$.

(a) Find estimates of the mean and the variance of the number of telephone units used in that three-month period in the town. [5]

$$\bar{x} = \frac{\Sigma fx}{250} = -0.16 \text{ and } \bar{t} = 10\bar{x} + 290$$

$$\rightarrow \bar{t} = -1.6 + 290 = 288.4$$

$$\sigma_x^2 = \frac{1075}{250} - (-0.16)^2 = 4.2744$$

(b) Suggest two ways in which the accuracy of your estimates could be improved. [2]

Examiner's Commentary

For help see Revise AS Study Guide section 4

Correct answer for $\bar{x}$ and has rearranged the coding formula.

Correct – check the order of magnitude makes sense.

Correct but candidate now forgets to multiply by 100 to find σ_t^2, **4/5 scored.**

The candidate is unable to answer – possible ways would be to take a larger sample and/or use the actual data rather than group it, **0/2 scored.**

GRADE BOOSTER

If $y = ax + b$, then

(i) $\bar{y} = a\bar{x} + b$

(ii) $\sigma_y^2 = a^2\sigma_x^2$.

Learn these results!

A grade candidate – mark scored 15/17

Examiner's Commentary

(1) The average weekly incomes, in £, of households in 11 regions of the UK are given below:

254, 251, 268, 297, 359, 289, 266, 261, 247, 259, 219

(a) Find the median and the lower and upper quartiles. [3]

219, 247, 251, 254, 259, 261, 266, 268, 289, 297, 359

$Q_2 = 261$; $Q_1 = 251$; $Q_3 = 289$

Correct – put the data in numerical order.

All correct, **3/3 scored.**

(b) On graph paper, draw a box plot to represent these data. [3]

200 210 220 230 240 250 260 270 280 290 300 310 320 330 340 350 360

All correct, **3/3 scored.**

There must be a scale

(c) Identify a possible outlier. [1]

A possible outlier is 359.

Correct, **1/1 scored.**

(d) Find the mean and standard deviation of these data. [6]

x	x^2
219	47961
247	61009
251	63001
254	64516
259	67081
261	68121
266	70756
268	71824
289	83521
297	88209
359	128881
2970	814880

$x = \frac{2970}{11}$

$= 270$

$sd = \sqrt{\left(\frac{814880}{11} - 270^2\right)}$

$= 34.35$ (2 d.p.)

All correct; the candidate has shown all the working which is a good idea – check your answer using the stats functions on your calculator, **6/6 scored.**

(e) Further investigation suggests that the £359 value should in fact be £309. Without further calculation say what effect this change would have on

(i) the standard deviation

(ii) the interquartile range,

explaining your answers. [4]

(i) The standard deviation will decrease since the spread of the data will decrease.

(ii) The interquartile range will decrease for the same reason.

Correct; the standard deviation takes all values into account.

Incorrect; the IQR is a measure of the spread of the 'middle half' of the data which is unaffected by the change; in general the IQR is not changed by extreme values, **2/4 scored.**

Exam practice questions

4.1 Representing data

1 The stem and leaf diagram below summarises the data giving the number of minutes, to the nearest minute, that a random sample of 65 trains from Guildford were late arriving at a main line station.

Minutes late	0 | 2 means 2	Totals
0	2 3 3 3 4 4 4 4 5 5 5 5 5 5	(14)
0	6 6 6 7 7 8 8 8 9	(9)
1	0 0 0 2 2 3 3 4 4 4 5	( )
1	6 6 7 7 8 8 8 9 9	( )
2	1 2 2 3 3 3 3 4	( )
2	6	( )
3	3 4 4 5	(4)
3	6 8	(2)
4	1 3	(2)
4	7 7 9	(3)
5	2 4	(2)

(a) Write down the missing values. [2]

(b) Find the median and the quartiles of these times. [3]

(c) Find the 65th percentile. [2]

(d) On graph paper construct a box plot for these data, showing the scale clearly. [3]

A random sample of trains arriving at the same main line station from Reading had a minimum value of 15 minutes late and a maximum of 30 minutes. The quartiles were 18, 22 and 27 minutes.

(e) On the same graph paper, using the same scale, construct a box plot for these data. [3]

(f) Compare and contrast the journeys from Guildford and Reading based on these data. [3]

4.2 Probability

1 **(a)** Show that, for any two events A and B:
$P(A \cup B) = P(A) + P(B) - P(A \cap B)$. [2]

(b) Express in words the meaning of $P(A \mid B)$. [1]

Given that A and B are independent events,

(c) express $P(A \cap B)$ in terms of $P(A)$ and $P(B)$, [1]

(d) show that A' and B are also independent. [3]

In a school, 60 pupils are studying one or more of the three subjects: Biology, Chemistry and Physics. Of these, 25 are studying Biology, 26 are studying Chemistry, 44 are studying Physics, 10 are studying Biology and Chemistry, 15 are studying Chemistry and Physics and 16 are studying Biology and Physics.

Answers on pages 76–80 Answers on pages 76–80 Answers on pages 76–80

(e) Write down the probability that a student, chosen at random from those studying Physics, is also studying Chemistry. [1]

(f) Determine whether or not the events 'studying Biology' and 'studying Chemistry' are independent. [2]

A student is chosen at random from all 60 students.

(g) Find the probability that the chosen student is studying all three subjects. [5]

2 A golfer has six different clubs in his golf bag, only one of which is correct for the shot about to be played. The probability that the golfer plays a good shot if the correct club is chosen is $\frac{3}{5}$ and the probability of a good shot if the incorrect club is chosen is $\frac{1}{4}$. The golfer chooses a club at random and plays the shot.

(a) Find the probability that a good shot is made. [4]

(b) The golfer plays a good shot. What is the probability that he chose the correct club? [3]

(c) Find the probability that an incorrect club was used given that a bad shot was made. [3]

(d) Comment on the model that the golfer chooses a club at random. [1]

3 The events A and B are independent and $P(A \mid B) = \frac{3}{4}$ and $P(B) = P(B' \cap A')$. By letting $P(B) = x$ and forming an equation in x, find

(a) $P(B)$, [5]

(b) $P(B' \cap A)$, [2]

(c) Write down $P(B \mid A)$. [1]

4 A bag contains 3 red, 4 white and 5 blue balls. Three balls are selected at random from the bag, without replacement. Find the probability that the three balls are of different colours.

4.3 Discrete random variables

1 A random number generator in a certain computer game produces values which can be modelled by the discrete random variable R whose probability function is given by

$$P(R = r) = kr! \quad r = 0, 1, 2, 3, 4$$

where k is a constant.

(a) Show that $k = \frac{1}{34}$. [2]

(b) Sketch the probability distribution of R. [2]

(c) Find $E(R)$ and $\text{Var}(R)$. [4]

Two independent values of R, R_1 and R_2, are generated.

(d) Find $P(R_1 = R_2)$. [3]

(e) Given that $R_1 = R_2$, find the probability that $R_1 = R_2 = 4$. [3]

Answers on pages 76–80 **Answers** on pages 76–80 **Answers** on pages 76–80

2 The random variable X has the following distribution:

x:	1	2	3
$P(X = x)$:	p	q	p

(a) Find $E(X)$. [5]

Given that $\text{Var}(X) = 0.75$,

(b) find the values of p and q. [1]

3 A spinner is made from the disc in the diagram and the random variable N represents the number that it falls on after being spun.

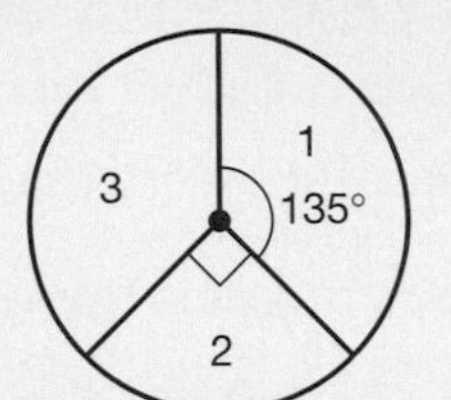

(a) Find the distribution of N. [2]

(b) Write down $E(N)$. [1]

(c) Find $\text{Var}(N)$. [3]

Sophie and Tom use the spinner to play a board game. Sophie's score is given by the random variable $2N - 1$ and Tom's score is given by the random variable $3N - 3$.

(d) Show that the mean score for each player is the same. [3]

(e) Find the variance of Sophie's score. [2]

Statistics 1

4.4 The Normal distribution

1 The random variable X is distributed normally with mean 5 and variance 4. Calculate

(a) $P(X = 64)$ [1]

(b) $P(X > 0)$ [3]

(c) $P(|X - 5| > 3)$. [5]

2 The random variable Y is normally distributed with mean μ and variance σ^2.
Given that $P(Y > 58.37) = 0.02$ and $P(Y < 40.85) = 0.01$, calculate μ and σ^2. [10]

3 A machine produces gaskets for engines. The engine manufacturer's specification is that the thickness of the gaskets should be between 5.45 mm and 5.55 mm, and the diameter should lie between 8.45 mm and 8.54 mm. The machine produces gaskets whose thicknesses are normally distributed with mean 5.5 mm and variance 0.0004 mm^2 and whose diameters are independently normally distributed with mean 8.54 mm and variance 0.0025 mm^2.
Find, to 1 decimal place, the percentage of gaskets produced which will not meet:

(a) the thickness specifications, [6]

(b) the diameter specifications. [6]

4.5 Correlation and regression

1 In order to assess the potential of ten prospective new pupils, a selective school sets a Mathematics test and a Verbal Reasoning (VR) test. Their results were as follows, where x represents their Mathematics score and y represents their VR score:

x	4	21	12	11	15	13	29	17	15	15
y	30	39	22	25	28	37	45	20	32	34

(a) Calculate the value of the product–moment correlation coefficient, r, between x and x, given that $\sum y^2 = 2696$, $\sum y^2 = 10288$ and $\sum xy = 5014$. [6]

A concerned parent of one of the prospective new pupils suggests that, in future, perhaps the school should set only one test in order to reduce the pressure on the children hoping to gain entry to the school. In response to this suggestion the Headteacher states that the school would like to retain both tests since they measure different abilities.

(b) Comment on the Headteacher's statement in the light of your answer to part **(a)**. [2]

2 Two branches of a small retail chain of grocery shops were situated close to each other in a small town. It was thought that sales in one may possibly be affecting the sales in the other and since no other shops nearby sold fresh vegetables it was decided to compare sales of fresh vegetables to see if this was true. Sales of vegetables were recorded weekly in pounds for seven weeks; the data collected are recorded in the following table:

Week	1	2	3	4	5	6	7
Shop A	380	402	370	365	410	392	385
Shop B	560	543	561	573	550	554	540

(a) Using a method of coding, calculate the product–moment correlation. [9]

(b) Comment on your result. [2]

3 A Head of Mathematics needs to make predictions about the final A level grade each of his students will achieve. To do this he decides to look at the marks obtained in mock examinations. In Mathematics he believes there is a linear relationship between the mark obtained in a mock examination and the final mark obtained. To investigate this he looks at the results from last year; the mock mark and final mark are given in the table below:

Mock mark x	18	26	28	34	36	42	48	52	54	60
Final mark y	54	64	54	62	68	70	76	66	76	74

(a) Draw a scatter diagram to illustrate these data. [2]

(b) Calculate the regression line in the form $y = a + bx$. [9]

(c) What final marks might be expected to be gained by students obtaining 16, 30, 50, 85 in their mock examination? [3]

(d) Comment on the validity of these predictions. [2]

Answers on pages 76–80 **Answers** on pages 76–80 **Answers** on pages 76–80

Answers

4.1 Representing data

examiner's tips

(1) (a) 11, 9, 8, 1.

(b) $\frac{1}{4}(65) = 16.25$ so Q_1 is 17th fig. i.e. 6 ← **If $\frac{1}{4}n$ is not an integer, round up.**

$\frac{1}{2}(65) = 32.5$ so Q_2 is 33rd fig. i.e. 14 ← **If $\frac{1}{2}n$ is not an integer, round up.**

$\frac{3}{4}(65) = 48.75$ so Q_3 is 49th fig. i.e. 23 ← **If $\frac{3}{4}n$ is not an integer, round up.**

(c) $\frac{65}{100}(65) = 42.25$ so P_{65} is 43rd fig. i.e. 19

(d) $Q_1 - 1.5(Q_3 - Q_1) = 6 - 1.5 \times 17 = -19.5$ ← **So no outliers at lower end.**

$Q_3 + 1.5(Q_3 - Q_1) = 23 + 1.5 \times 17 = 48.5$ ← **So 49, 52, 54 are all outliers.**

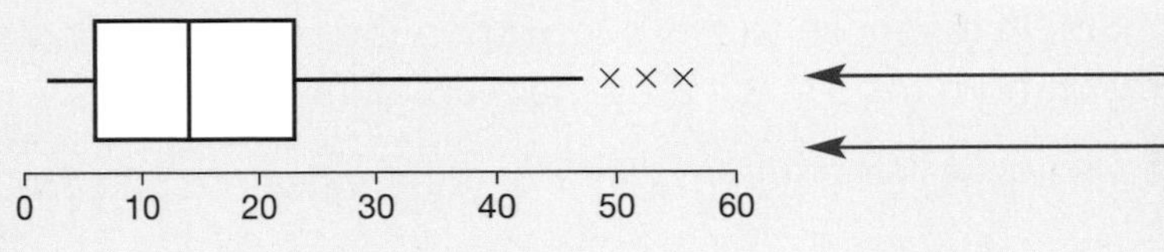

← **LH whisker starts at lowest value.**

← **RH whisker stops at 48.5 with the three outliers marked individually.**

(e) $Q_1 - 1.5(Q_3 - Q_1) = 18 - 1.5 \times 9 = 4.5$ ← **So no outliers at lower end.**

$Q_3 + 1.5(Q_3 - Q_1) = 27 + 1.5 \times 9 = 40.5$ ← **So no outliers at upper end.**

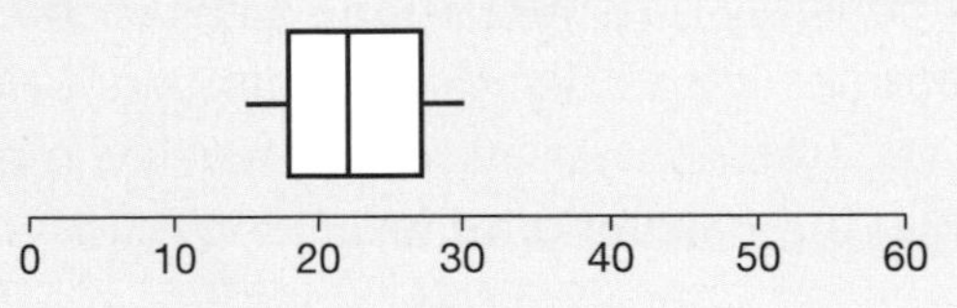

(f) Both likely to be late but the lateness for the Guildford service is more inconsistent and could be quite severe. The spread for the Reading service is smaller and hence it may be easier to plan for.

4.2 Probability

(1) (a) $n(A \cup B) = n(A) + n(B) - n(A \cap B)$ ← **Since A ∩ B is counted twice.**

Dividing through by $n(\varepsilon)$ gives the result.

(b) Probability of A given B.

(c) $P(A \cap B) = P(A).P(B)$.

(d) $P(A').P(B) = (1 - P(A)).P(B)$ ← **We must show that P(A')P(B) = P(A' ∩ B).**

$= P(B) - P(A).P(B)$

$= P(B) - P(A \cap B)$ ← **Using the independence of A and B.**

$= P(A' \cap B)$ hence A' and B are independent.

(e) $\frac{15}{44}$

(f) $P(B) = 25/60$; $P(B \mid C) = 10/26$ as these are different B and C are not independent. ← **A more intuitive definition of the independence of B and C is: P(B) = P(B/C) i.e. the outcome of B is not affected by the outcome of C.**

Statistics 1

(g)

(60)
$B(25)$ $C(26)$ $P(44)$
$10-x$ $x-1$ $1+x$ x $16-x$ $15-x$ $13+x$ O

Using a Venn Diagram is the right approach – always work out from the centre when entering the data.

Suppose there are x pupils doing all three then $44 + 10 - x + x - 1 + x + 1 = 60$, so $x = 6$; hence probability is $\frac{6}{60} = \frac{1}{10}$.

(2) (a) $P(\text{correct club}) = \frac{1}{6}$ and $P(\text{incorrect club}) = \frac{5}{6}$ ← Since there are 6 clubs to choose from.

$P(\text{good shot}) = (\frac{1}{6} \times \frac{3}{5}) + (\frac{5}{6} \times \frac{1}{4}) = \frac{37}{120}$ ← A tree diagram could have been used.

(b) $P(\text{correct club given a good shot}) = \dfrac{(\frac{1}{6} \times \frac{3}{5})}{(\frac{37}{120})} = \frac{12}{37}$

(c) $P(\text{bad shot}) = (\frac{1}{6} \times \frac{2}{5}) + (\frac{5}{6} \times \frac{3}{4}) = \frac{83}{100}$ ← Or we could have used $(1 - \frac{37}{120})$.

$P(\text{incorrect club given a bad shot}) = \dfrac{(\frac{5}{6} \times \frac{3}{4})}{(\frac{83}{120})}$ ← This is a conditional probability.

$= \frac{75}{83}$

(d) This is totally unrealistic.

(3) (a) $P(A) = P(A \mid B) = \frac{3}{4}$ since A and B are independent and A' and B' are also independent ← As A and B do not affect each other $P(A) = P(A \mid B)$.

so $P(B' \cap A') = P(B').P(A')$

So, $x = (1 - x).\frac{1}{4} \rightarrow x = \frac{1}{5}$ i.e. $P(B) = \frac{1}{5}$

(b) $P(B' \cap A) = P(B').P(A) = \frac{4}{5} \times \frac{3}{4} = \frac{3}{5}$ ← As B and A are independent.

(c) $P(B \mid A) = P(B) = \frac{1}{5}$ ← As B and A are independent.

(4) Prob $= \frac{3}{12} \times \frac{4}{11} \times \frac{5}{10} \times 3! = \frac{3}{11}$ ← There are 3! ways of choosing a red, white and blue ball. The denominators of the fractions decrease by 1 each time as the selection is without replacement.

4.3 Discrete random variables

(1) (a) $k(0! + 1! + 2! + 3! + 4!) = 1$ ← The sum of the probabilities is 1.

$k(1 + 1 + 2 + 6 + 24) = 1 \rightarrow k = \frac{1}{34}$

(b)

r	0	1	2	3	4
$P(R = r)$	$\frac{1}{34}$	$\frac{1}{34}$	$\frac{2}{34}$	$\frac{6}{34}$	$\frac{24}{34}$

← This is the probability distribution of R.

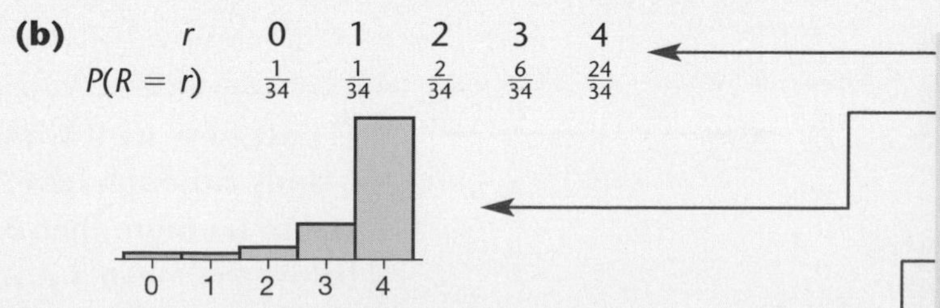

← This is a histogram showing the distribution of R – note that the total area is 1.

(c) $E(R) = 0 \times \frac{1}{34} + 1 \times \frac{1}{34} + 2 \times \frac{2}{34} + 3 \times \frac{6}{34} + 4 \times \frac{24}{34}$

$= \frac{119}{34} = 3\frac{17}{34} = 3.5$

$\text{Var}(R) = E(R^2) - [E(R)]^2$ ← Note that it is possible to find both $E(R)$ and $\text{Var}(R)$ on a calculator by treating the probabilities as frequencies – use $\bar{x}$ for $E(R)$ and $(\sigma_x)^2$ for $\text{Var}(R)$.

$= 0^2 \times \frac{1}{34} + 1^2 \times \frac{1}{34} + 2^2 \times \frac{2}{34} + 3^2 \times \frac{6}{34} + 4^2 \times \frac{24}{34} - (3.5)^2$

$= 0.897$ (3 s.f.)

(d) $P(0, 0) + P(1, 1) + P(2, 2) + P(3, 3) + P(4, 4)$

$= \{P(0)\}^2 + \{P(1)\}^2 + \{P(2)\}^2 + \{P(3)\}^2 + \{P(4)\}^2$ ← Since R_1 and R_2 are independent.

$= 0.535$ (3 s.f.)

(e) $P(R_1 = R_2 = 4) = (\frac{24}{34})^2 = 0.498$ ← A conditional probability is required.

$$\frac{P(R_1 = R_2 = 4)}{P(R_1 = R_2)}$$

$$= \frac{0.49826\ldots}{0.5346} = 0.932 \text{ (3 s.f.)}$$ ← Note that we have to work to at least 4 s.f. to be sure of obtaining an answer which is correct to 3 s.f.

(2) (a) $E(X) = p + 2q + 3p = 4p + 2q$

Also $2p + q = 1$, ← Since the sum of the probabilities is 1.

So $E(X) = 2(2p + q) = 2$.

(b) $\text{Var}(X) = E(X^2) - [E(X)]^2$

$0.75 = p + 4q + 9p - 2^2$

$\rightarrow \quad 10p + 4q = \frac{19}{4}$

And $\quad 2p + q = 1$;

Solving gives $p = \frac{3}{8}$ and $q = \frac{1}{4}$.

(3) (a)

n:	1	2	3
$P(N = n)$:	$\frac{3}{8}$	$\frac{2}{8}$	$\frac{3}{8}$.

← The probabilities are proportional to the angles by symmetry.

(b) $E(N) = 2$. ← By symmetry.

(c) $\text{Var}(N) = \frac{1}{8}(3 + 8 + 27) - 2^2 = 0.75$. ← $\text{Var}(N) = E(N^2) - [E(N)]^2$.

(d) $E(2N - 1) = 2E(N) - 1 = 3$ ← $E[aN + b] = aE[N] + b$.

$E(3N - 3) = 3E(N) - 3 = 3$.

(e) $\text{Var}(2N - 1) = 4\text{Var}(N) = 3$. ← $\text{Var}[aN + b] = a^2\text{Var}[N]$.

4.4 The Normal distribution

(1) (a) $P(X = 64) = 0$ ← Since there is no area – a surprising result.

(b) $P(X > 0) = P\left(Z > \frac{(0 - 5)}{2}\right)$ ← Standardising the variable.

$= P(Z < 2.5)$ ← By symmetry – draw a diagram.

$= \Phi(2.5)$

$= 0.9938$

(c) $P(|X - 5| > 3) = P(X > 8 \text{ or } X < 2)$ ← It is best to write it without the $|\,|$ signs although here, as $\mu = 5$, X has to be more than 3 above or 3 below the mean, i.e. more than 1.5 SD above or below the mean, i.e. prob $= 2(1 - \Phi(1.5))$.

$= P(X > 8) + P(X < 2)$

$= 1 - P(X < 8) + P(X < 2)$

$= 1 - P\left(Z < \frac{(8 - 5)}{2}\right) + P\left(Z < \frac{(2 - 5)}{2}\right)$

$= 1 - P(Z < 1.5) + P(Z < -1.5)$

$= 1 - P(Z < 1.5) + 1 - P(Z < 1.5)$

$= 2 - 2\,\Phi(1.5)$

$= 0.1336$

(2)

$$P(Y > 58.37) = 0.02$$
$$1 - P(Y < 58.37) = 0.02$$ ← From a diagram.
$$P(Y < 58.37) = 0.98$$
$$P\left(Z < \frac{(58.37 - \mu)}{\sigma}\right) = 0.98$$ ← Standardising the variable.
$$\Phi\left(\frac{(58.37 - \mu)}{\sigma}\right) = 0.98$$
$$\left(\frac{(58.37 - \mu)}{\sigma}\right) = 2.05$$
$$\mu + 2.05\sigma = 58.37 \quad (1)$$ ← Rearranging.
$$P(Y < 40.85) = 0.01$$
$$P\left(Z, \frac{(40.85 - \mu)}{\sigma}\right) = 0.01$$ ← Standardising the variable.
$$\Phi\left(\frac{(40.85 - \mu)}{\sigma}\right) = 0.01$$
$$\Phi\left\{-\left(\frac{(40.85 - \mu)}{\sigma}\right)\right\} = 0.99$$ ← We cannot read back from the tables as 0.01 < 0.5; using the symmetry of the graph.
$$\mu - 2.33\sigma = 40.85 \quad (2)$$
$$4.38\sigma = 17.52$$ ← (1) − (2) to eliminate μ.
$$\sigma = 4 \text{ i.e. } \sigma^2 = 16$$
$$\text{so } \mu - 9.32 = 40.85$$ ← Substitute for σ in (2).
$$\mu = 50.17$$

(3) (a) $P(5.45 < T < 5.55) = P\left(-\frac{0.05}{0.02} < Z < \frac{0.05}{0.02}\right)$ ← First find the probability that they do meet the specification.

$= P(-2.5 < Z < 2.5) = \Phi(2.5) - \Phi(-2.5)$

$= \Phi(2.5) - \{1 - \Phi(2.5)\} = 2\Phi(2.5) - 1 = 0.9876$ ← Using the symmetry of the graph.

Hence, answer is $(1 - 0.9876) \times 100\% - 1.2\%$ (to 1 d.p.).

(b) $P(8.45 < D < 8.54) = P\left(-\frac{0.09}{0.05} < Z < 0\right)$ ← First find the probability that they do meet the specification.

$= P(-1.8 < Z < 0) = \Phi(0) - \Phi(-1.8)$

$= 0.5 - 1 + \Phi(1.8) = 0.4641$ ← Using the symmetry of the graph.

Hence, answer is $(1 - 0.4641) \times 100\% = 53.6\%$ (to 1 d.p.).

4.5 Correlation and regression

(1) (a) $\Sigma x = 152 \quad \Sigma y = 312 \quad \Sigma x^2 = 2696 \quad \Sigma y^2 = 10288 \quad \Sigma xy = 5014.$

$$S_{xx} = 2696 - \frac{152^2}{10} = 385.6$$ ← It is best to set out the calculation like this.

$$S_{yy} = 10288 - \frac{312^2}{10} = 553.6$$

$$S_{xy} = 5014 - \frac{152 \times 312}{10} = 271.6$$

$$r = \frac{S_{xy}}{\sqrt{S_{xx}S_{yy}}} = \frac{271.6}{\sqrt{(385.6 \times 553.6)}} = 0.588 \text{ (3 s.f.)}$$ ← You can use a calculator to check this but you risk losing all the marks if you just put down the answer, and it's wrong.

(b) This value of r suggests a high positive correlation between the two tests, i.e. one test should do.

Statistics 1

(2) (a) Let $a = A - 370 \quad b = B - 540$

a	b	a^2	b^2	ab
10	20	100	400	200
32	3	1024	9	96
0	24	0	576	0
−5	33	25	1089	−165
40	10	1600	100	400
22	14	484	196	308
15	0	225	0	0
114	104	3458	2370	839

$$S_{aa} = 3458 - \frac{114^2}{7} = 1601.43$$

$$S_{bb} = 2370 - \frac{104^2}{7} = 824.86$$

$$S_{ab} = 839 - \frac{114 \times 104}{7} = -854.71$$

$$r = \frac{S_{ab}}{\sqrt{(S_{aa}S_{bb})}} = \frac{-854.71}{\sqrt{(1601.43 \times 824.86)}} \; -0.744 \text{ (3 s.f.)}.$$

You can use a calculator to check this.

(b) This value of r indicates a high negative correlation between the sales in each shop – i.e. if they're up in one they're likely to be down in the other and vice versa.

(3) (a)

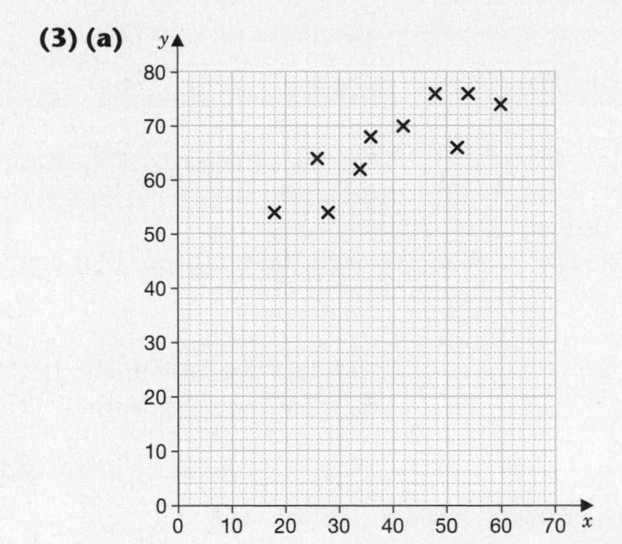

(b) $\Sigma x = 398 \; \Sigma y = 664 \; \Sigma x^2 = 17524 \; \Sigma y^2 = 44680 \; \Sigma xy = 27268$

$$S_{xx} = 17524 - \frac{398^2}{10} = 1683.6$$

$$S_{xy} = 27268 - \frac{398 \times 664}{10} = 840.8$$

It is best to set out the calculation like this.

$$b = \frac{840.8}{1683.6} = 0.4994 = 0.5 \text{ (2 s.f.)}$$

It is best to set out the calculation like this.

$$a = \bar{y} - b\bar{x} = 66.4 - 0.5 \times 39.8 = 46.5$$

i.e. $y = 46.5 + 0.5x$.

(c) and **(d)** When $x = 30$, $y = 61.5$
When $x = 50$, $y = 71.5$
When $x = 16$, $y = 54.5$

However, $x = 85$ cannot be used, with any degree of confidence, to predict as it is well outside the range of the data.

Both 30 and 50 are well within the range of the data collected; so these predictions are fairly safe.

16 is just outside the range of the data so this value is fairly reliable.

Questions with model answers

C grade candidate – mark scored 9/12

(1) The highest common factor (HCF) of two numbers A and B in which $A > B$ can be found using the following algorithm.

Step 1: Input A and B.

Step 2: Take B from A.

Step 3: If $A > B$ then go back to step 2.

Step 4: If $A = B$ then print A and stop.

Step 5: Reverse A and B and return to step 2.

(a) Demonstrate this algorithm when $A = 189$ and $B = 153$. **[8]**

A:	189	36	153	117	81	45	9	36	27	18	9
B:	153	153	36	36	36	36	36	9	9	9	9
Steps:	1	2,3,4	5	2,3	2,3	2,3	2,3,4	5	2,3	2,3	2,3,4

(b) Demonstrate what happens if A and B are interchanged and explain why it is necessary to have the initial condition $A > B$. **[4]**

A:	153	-36	189	225
B:	189	189	-36	-36
Steps:		2	5	2

Because after step 2 A would be a negative number.

Examiner's Commentary

This is correct – candidates are strongly advised to write out each step as shown. This is not only so that the examiner can see the steps but so that the work can be checked. However, crucially, the printout is not given! **7/8 scored**.

This is so, but the process would not fall down there, because step 3 is satisfied, step 4 is not and so step 5 would take place, resulting in A being positive and B negative; in this case $A = 189$ and $B = -36$. Now successive subtractions of B from A will only increase the value of A and an everlasting cycle will be initiated. The candidate should be aware that more is required for full marks, **2/4 scored**.

GRADE BOOSTER

When demonstrating an algorithm, it is essential that full working is shown and a full explanation of how it works is given.

A grade candidate – mark scored 15/18

(1) A school is making two types of 'music lover's' stationary packs to raise funds for a choir tour. Pack A contains 100 sheets of paper, 60 envelopes and 40 notelets. Pack B contains 50 sheets of paper, 50 envelopes and 60 notelets.

The school has 5000 sheets of paper, 3300 envelopes and 3000 notelets available for the packs.

The school will make a profit of £6 on each pack A sold and £8 on each pack B sold.

Let x and y be the number of pack A and pack B produced respectively

(a) Formulate this situation as a linear programming problem. [7]

$P = 6x + 8y$

$100x + 50y \le 5000$ so $2x + y \le 100$

$60x + 50y \le 3300$ so $6x + 5y \le 330$

$40x + 60y \le 3000$ so $2x + 3y \le 150$

(b) On graph paper display your inequalities, labelling the feasible region R. [6]

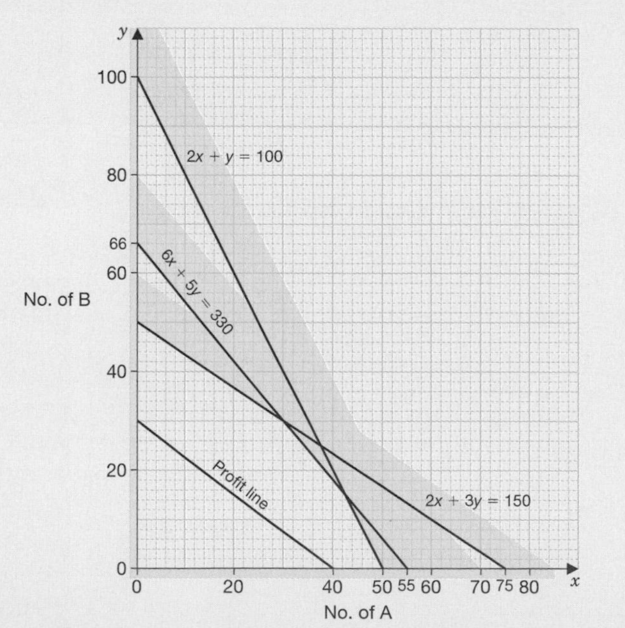

(c) Determine the number of each pack that the school should produce and the profit they make, making your method clear. [5]

Say P = 240 then (0, 30) (40 0)

profit line drawn on graph

By ruler method, max is where $2x + 3y = 150$ and $6x + 5y = 330$

$6x + 9y = 450$

$6x + 5y = 330$

$4y = 120$

So $x = 30$ and $y = 30$.

$P = 6(30) + 8(30) = £420$

Examiner's Commentary

The candidate has managed to get all of the inequalities correct and has even simplified them to make the next part (drawing the lines) easier. However the candidate has not said that they are trying to **maximise P**, and has forgotten the non-zero constraints x, y ⩾ 0, **5/7 scored**.

The candidate has drawn (and labelled) the lines correctly and the shading is correct, but the candidate has not labelled the feasible region, **5/6 scored**.

There are two methods of finding the optimal point – the vertex method or the profit line method. This candidate has chosen the profit line method. The working is all correct and there is enough written down to make the method clear, **5/5 scored**.

Exam practice questions

5.1 Sorts, searches and packing

1 Sort the list below into ascending order using a bubble sort. You should show each exchange for the first pass, then the result of each subsequent pass.

25 84 67 16 35 79 27 56 44 [5]

5.2 Minimum spanning trees (minimum connectors)

1

	A	*B*	*C*	*D*	*E*	*F*
A	–	53	17	68	56	49
B	53	–	30	36	61	24
C	17	30	–	53	40	67
D	68	36	53	–	45	57
E	56	61	40	45	–	38
F	49	24	67	57	38	–

The table shows the distance in km between six towns *A, B, C, D, E* and *F*.

(a) Use Prim's algorithm, starting at *A*, to find a minimum spanning tree. You must make your method clear by listing the order in which you included the arcs. [4]

(b) Draw your minimum spanning tree and state its length. [3]

2 **(a)** Use Kruskal's algorithm to find a minimum connector for the network opposite. Make your method clear. [4]

(b) State, with a reason, whether your minimum connector is unique. [2]

N = the number of nodes in the network.

E = the number of edges in a minimum connector.

(c) Write down the relationship between N and E. [1]

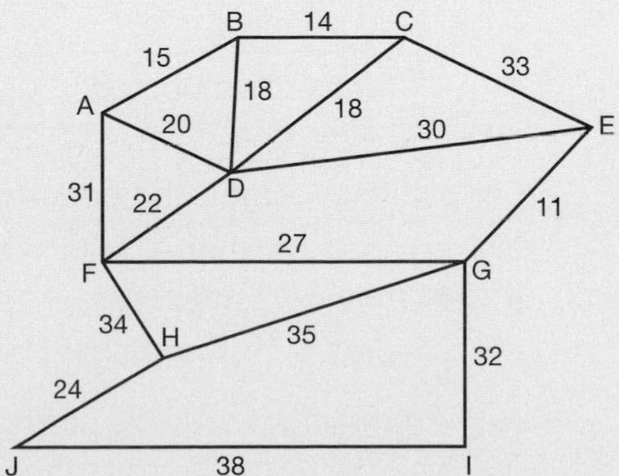

Answers on pages 86–88 Answers on pages 86–88 Answers on pages 86–88

5.3 Dijkstra's algorithm

1 The diagram represents the roads joining eight villages, labelled A–H. The numbers give distances in km.

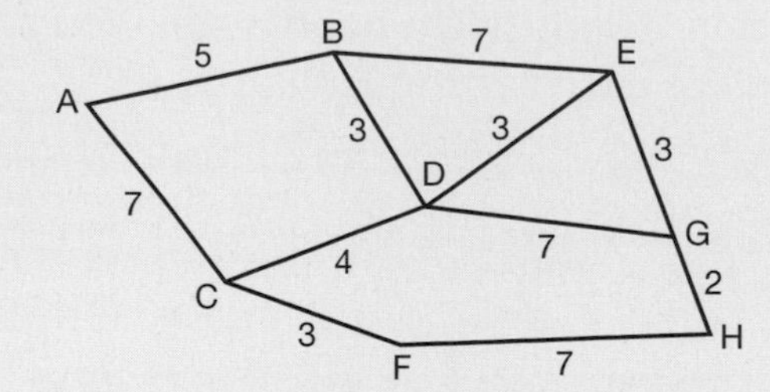

(a) Use Dijkstra's algorithm to find a shortest route from A to H. Explain the method carefully and show all your working. Give the shortest route and its length. **[5]**

(b) Find the time that John takes to travel by the quickest route if he drives on average 40 kilometres per hour. **[2]**

(c) One particular day John finds that the road from B to D is flooded and there is a fallen tree on the road from C to D. He knows that each of these obstructions adds 5 minutes to the journey times. Find the shortest time that John will take on the day with these obstructions and give the route that he should take. **[3]**

5.4 Linear programming

1 A company has to post small pieces of equipment, A and B, to a new customer. To do this the company uses special boxes, I and II. Each box will hold a mixture of equipment A and B.

The numbers of each type of equipment each box can hold, the postage cost and the total number of pieces of equipment that need to be posted are given in the table below.

	Equipment A	Equipment B	Postage cost (£)
Box I	220	80	1.50
Box II	110	70	2.00
Totals to be posted	12 100	5600	

For packing reasons the company decides to use at least half as many box II as box I.

The company wishes to minimise the cost of posting the boxes. The numbers of box I and box II used must be integers.

Let x and y be the numbers of box I and box II used respectively.

(a) Formulate the above problem as a linear programming problem. **[6]**

(b) Determine how many of each box the company should use to minimise the postal cost and state what this cost is. **[10]**

Answers on pages 86–88 Answers on pages 86–88 Answers on pages 86–88

5.5 Critical path analysis

1 The table shows activities involved in a building project, with their duration (in hours) and immediate predecessors.

Activity	A	B	C	D	E	F	G	H
Immediate predecessors	–	–	B	A, C	A, B	E	D, E	G
Duration	5	6	2	5	6	7	6	3

(a) Complete the activity network. [4]

(b) Find the early and late times by performing a forward and backward pass. [4]

(c) Give the critical path and the minimum time to completion. [2]

Answers on pages 86–88 Answers on pages 86–88 Answers on pages 86–88

Answers

5.1 Sorts, searches and packing

(1) Sorting from left to right.

First pass

25	**84**	**67**	16	35	79	27	56	44
25	67	**84**	**16**	35	79	27	56	44
25	67	16	**84**	**35**	79	27	56	44
25	67	16	35	**84**	**79**	27	56	44
25	67	16	35	79	**84**	**27**	56	44
25	67	16	35	79	27	**84**	**56**	44
25	67	16	35	79	27	56	**84**	**44**
25	67	16	35	79	27	56	44	84

The subsequent passes give:

25	16	35	67	27	56	44	79	84
16	25	35	27	56	44	67	79	84
16	25	27	35	44	56	67	79	84

The next pass has no changes so the list is in order.

examiner's tips

You must be consistent and keep moving in the same direction until you get to the end of the list. Candidates often lose (or change) numbers when sorting. Keeping your numbers in columns helps prevent this first error, and checking the final list with the original list helps prevent the second.

5.2 Minimum Spanning Trees (Minimum Connectors)

(1) (a)

	1	3	2	5	6	4
	A	*B*	*C*	*D*	*E*	*F*
A	–	53	17	68	56	49
B	53	–	(30)	36	61	24
C	(17)	30	–	53	40	67
D	68	(36)	53	–	45	57
E	56	61	40	45	–	(38)
F	49	(24)	67	57	38	–

Order of arc selection: AC, BC, BF, BD, EF

Make sure that the MST values in your table are still legible after you have deleted each row – or you'll have difficulty in finding the length of your tree in part (b)! Writing the order in which you introduced each vertex gives the examiner another clue to confirm that you are using Prim correctly.

Do make sure that you scan ALL the columns you have highlighted EACH time – and not just the last one. You may miss a small value in a column you highlighted a while ago.

(b) e.g.

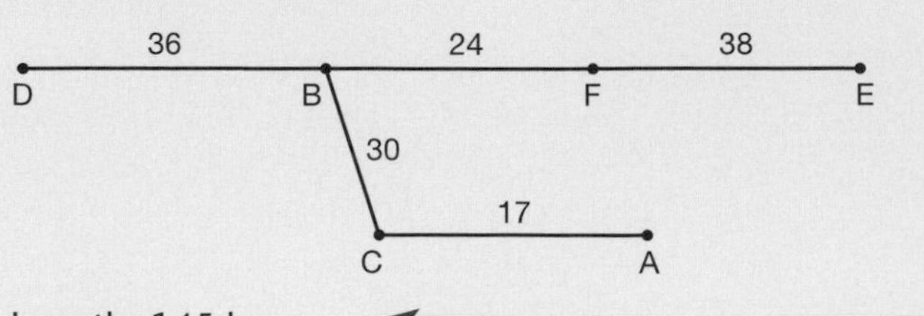

Length: 145 km

Don't forget units.

(2) (a) EG, BC, AB, (BD **or** CD – and the other rejected), reject AD, DF, HJ, FG, reject DE, reject AF, GI, reject CE, FH.

> **Remember that you don't have to be told where to start Kruskal's algorithm – you must start with the shortest edge – EG. Kruskal's algorithm often 'jumps' around the network in its hunt for the next smallest edge. Keep alert! Always show the examiner that you have considered the next arc, but then reject it if it forms a cycle.**

> **Do check that you have minimum connector before you stop! You must make sure that the tree is all linked up.**

(b) No it is not unique; there is a choice of using either BD or CD, hence there will be two minimum connectors, one using BD and the other using CD.

> **Do make sure you answer the question. You have to give a reason as well as stating if is unique or not. Just a statement about uniqueness alone is unlikely to gain any marks.**

(c) N = E − 1

> **If you find it hard to go directly to algebra – or you can't work out which side the 1 should be on, do use the minimum connector you have just found to help you work out the link between E and N. If necessary draw another tree to see if you got it right.**

5.3 Dijkstra's algorithm

(1) (a)

> **The diagram shows Dijkstra's algorithm applied using one notation system. You may have learnt another system, but the result will be the same! Fill in the boxes carefully, checking each step to ensure that you have it right.**
>
> **Be careful when crossing out numbers – or better yet don't cross them out at all! The examiner needs to be able to read all of these numbers. The examiner will also be checking to see if you have the working values in the correct order. These numbers are the clearest indication that you are using the algorithm correctly.**

The shortest distance is via ABDEGH and is 16 km.

(b) The time taken is 24 minutes.

> **Change each length to time, or take the total distance of 16 km at 40 km per hour, giving 24 minutes.**

(c) The route will now be round either perimeter and will take 25.5 minutes.

> **Either change all distances to times, adding 5 minutes to the routes CD and BD, or add a 'distance' of 7.5 km, which is the distance that would be travelled in 5 minutes at 40 km per hour. Then rework the algorithm.**

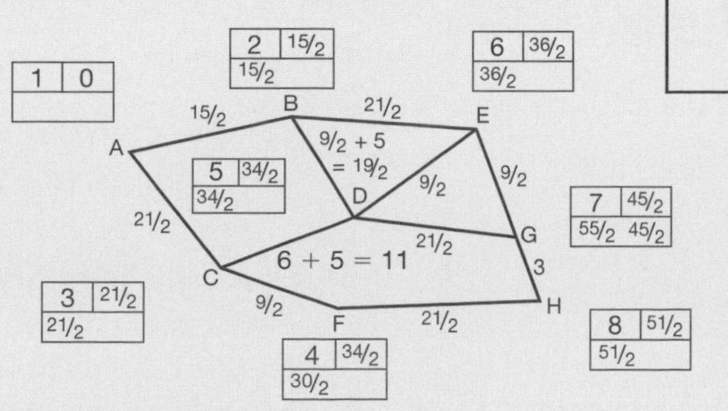

5.4 Linear programming

(1) (a) Minimise $C = 1.50x + 2y$

Subject to $220x + 110y \geq 12\,100 \Rightarrow 2x + y \geq 110$

$80x + 70y \geq 5\,600 \Rightarrow 8x + 7y \geq 560$

$x \leq 2y$

$x, y \geq 0$

> This type of inequality is always tricky. Sometimes experimenting with numbers helps – e.g. try $x = 7$ then y could be 4, 5, 6 etc. this helps decide which way the inequality goes and which side the 2 goes.

(b)

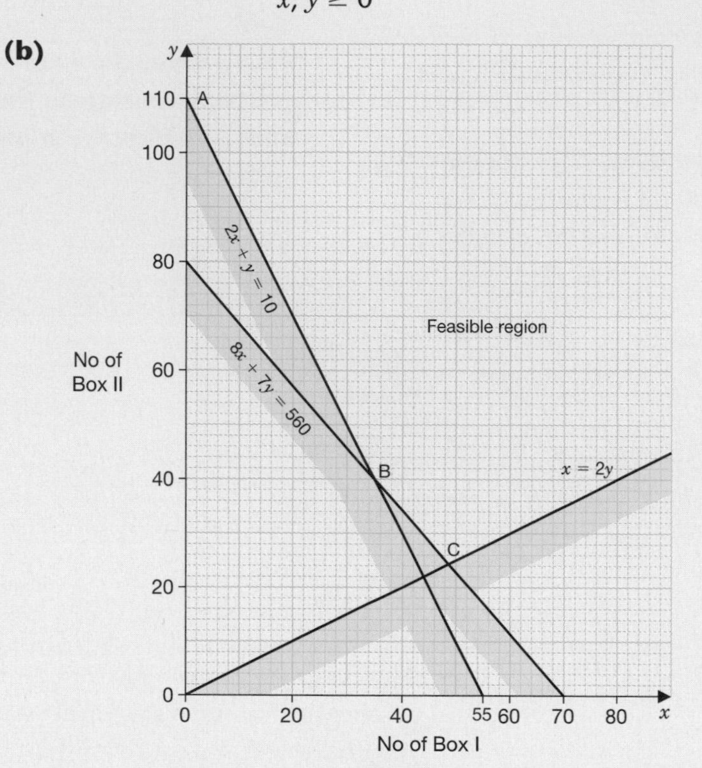

Vertex method A(0, 110) Cost = £220

B(35, 40) Cost = £132.50

C(48 16/23, 24 8/23) Cost = £121.74

But this minimum is not possible since x and y are not integers.

Test integer points near to C,

(48,24), (48,25) and (49, 24) lie outside the feasible region.

(48, 26) Cost = £124

(49, 25) Cost = £123.50

The company should use 49 of box I and 25 of box II at a postal cost of £123.50.

> All integer points near to the 'optimal vertex' need testing to see firstly if they lie in the feasible region, and secondly if they yield the optimal value. Do make your final answer clear – in terms of the original problem.

5.5 Critical path analysis

(1) (a) and **(b)**

6 | 6 8 | 8 13 | 13 19 | 19

C(2) D(5) G(6)

B(6) H(3)

A(5) E(6) F(7)

6 | 7 12 | 13 22 | 22

> Complete the diagram carefully with earliest and latest times.

> Check the line where the difference between latest time of one and the earliest time of the one before equals the length of the activity.

(c) Critical path B, C, D, G, H.

Minimum time: 22 hours.

AS Mock Exam 1

Centre number	
Candidate number	
Surname and initials	

Examining Group

Maths Core 1

Time: 1 hour 30 minutes Maximum marks: 75

Instructions

Answer **all** questions in the spaces provided. Show all steps in your working.
The marks allocated for each question are shown in brackets.

Grading
Boundary for A grade 60/75
Boundary for C grade 45/75

N.B. Calculators should NOT be used.

1 Calculate $\sum_{r=1}^{30}(4 - 3r)$. [3]

2 Find $\int\left(2x - \frac{4}{\sqrt{x}}\right)dx$. [4]

3 **(a)** Express $\sqrt{48}$ in the form $a\sqrt{3}$, where a is an integer. [1]

(b) Express $(2 - 3\sqrt{3})^2$ in the form $b + c\sqrt{3}$, where b and c are integers. [3]

4 The points P and Q have coordinates (2, 6) and (8, −4) respectively.

(a) Find the coordinates of M, the mid-point of PQ. [1]

(b) Find an equation for the perpendicular bisector of PQ, giving your answer in the form $ax + by + c = 0$, where a, b and c are integers. [5]

5 The diagram shows a sketch of part of the curve with equation $y = f(x)$.
The curve crosses the coordinate axes at the points (0, 3), (1, 0) and (3, 0).
The minimum point on the curve is (2, −1).
On separate diagrams, sketch the curve with equation.

(a) $y = f(x + 2)$, [3]

(b) $y = f(\frac{1}{2}x)$. [3]

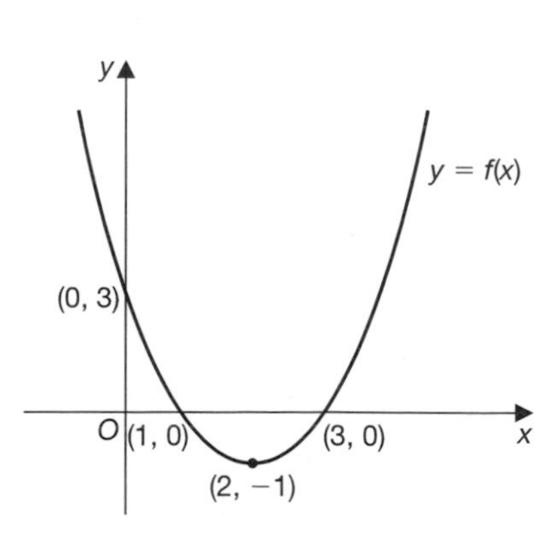

6 **(a)** Solve the simultaneous equations

$$x^2 + y - 10 = 0$$
$$x - y - 2 = 0.$$ [6]

(b) Hence, or otherwise, find the set of values of x for which

$$10 - x^2 > x - 2.$$ [3]

7 The first and third terms of an arithmetic series are 6 and $4y$ respectively.
Find, simplifying your answers, expressions in terms of y for

(a) the common difference of the series, [3]

(b) the fifth term of the series, [3]

(c) the sum of the first eleven terms of the series. [3]

8 Given that

$$x^2 + 8x + 25 \equiv (x + p)^2 + q,$$

where p and q are constants,

(a) find the value of p and the value of q. [3]

(b) Hence show that the equation $x^2 + 8x + 13 = -12$ has no real roots.

The equation $x^2 + 8x + R = 0$ has equal roots. [2]

(c) Find the value of R. [2]

(d) For this value of R, sketch the graph of $y = x^2 + 8x + R$, showing the coordinates of any points at which the graph meets the axes. [4]

9 The curve C has equation $y = f(x)$ and the point $A(2, 3)$ lies on C.

Given that

$$f'(x) = 3x^2 - 6x + 2,$$

(a) find $f(x)$. [4]

(b) Verify that the point $(1, 3)$ lies on C. [1]

The point B also lies on C, and the tangent to C at B is parallel to the tangent to C at A.

(c) Find the x-coordinate of B. [5]

10 The curve K has equation

$$y = 2 + \frac{3}{x}, x \neq 0.$$

The curve intersects the x-axis at the point A.

(a) Find the coordinates of the point A. [2]

(b) Find an equation of the normal to K at the point A. [6]

(c) Show that the normal intersects the curve again at the point with coordinates $(\frac{8}{3}, \frac{25}{8})$. [5]

AS Mock Exam 2

Centre number	
Candidate number	
Surname and initials	

Examining Group

Maths Core 2

Time: 1 hour 30 minutes Maximum marks: 75

Instructions

Answer **all** questions in the spaces provided. Show all steps in your working.
The marks allocated for each question are shown in brackets.

Grading
Boundary for A grade 60/75
Boundary for C grade 45/75

---

1 The circle C has centre $(2, -1)$ and passes through the point $(4, -3)$.
Find an equation for C. **[4]**

2 Find the first three terms, in descending powers of x, of the expansion of $(3 - 2x)^5$. **[4]**

3 **(a)** Given that $y = \sqrt{x^2 + 4}$, complete the following table, giving your answers to 3 decimal places.

x	1	2	3	4	5
y	2.236	2.828	3.606		

[2]

(b) Using the trapezium rule, with four strips, estimate the value of

$$\int_1^5 \sqrt{x^2 + 4}\,dx,$$

giving your answer to 2 decimal places. **[4]**

4 Solve, for $0° \leqslant x \leqslant 360°$,

$3\sin^2 x - \cos^2 x = 2\cos x + 3.$ **[7]**

5 A tray consists of two identical sectors, OPS and OQR, and two congruent triangles, OPQ and OSR, as shown in the diagram.

$\angle POS = 60°$ and OP = 24 cm.

Find, in exact form,

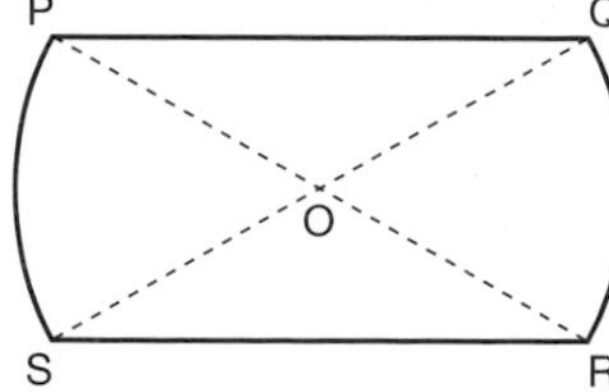

(a) the perimeter of the tray, **[3]**

(b) the area of the tray. **[4]**

6 A savings account pays 0.8% interest on the amount in the account at the end of each month.

(a) Peter invests £500 in this account.

Find, to the nearest penny, the amount in the account after the payment of interest at the end of the first year. [3]

(b) Abigail pays £200 into this account at the start of each month. Find, to the nearest penny, the amount in her account after the payment of interest at the end of a three-year period. [7]

7 $f(x) \equiv 2x^3 + px^2 + qx - 18$

Given that $(x + 3)$ and $(2x + 3)$ are factors of $f(x)$,

(a) find the values of p and q. [6]

(b) Hence solve $f(x) = 0$. [3]

(c) State the number of solutions of the equation

$$2\sin^3 x + 5\sin^2 x = 9\sin x + 18$$ [3]

8

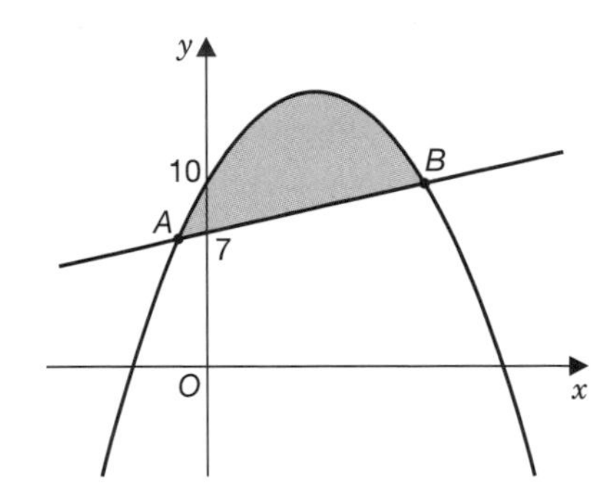

The line with equation $y = x + 7$ cuts the curve with equation $y = 10 + 3x - x^2$ at the points A and B, as shown in the diagram.

(a) Find the coordinates of the points A and B. [5]

(b) Find the area of the shaded region between the curve and the line, as shown in the diagram. [7]

9 A solid right circular cylinder has a fixed volume of 1000 cm^3. The base radius of the cylinder is x cm and the total surface area of the cylinder is A cm^2.

(a) Show that $A = 2\pi x^2 + \dfrac{2000}{x}$. [5]

(b) Use calculus to prove that A is a minimum when $x^3 = \dfrac{500}{\pi}$. Explain clearly how you know that this value of x gives the minimum area. [8]

AS Mock Exam 1 Answers

Method (M) marks for 'knowing a method and attempting to apply it'.
Accuracy (A) marks can only be awarded if the relevant (M) mark(s) have been earned.
(B) marks are independent of method marks.

(1) $S_{30} = \frac{30}{2}\{2 + (29 \times -3)\}$ M1A1

$= 15 \times -85$

$= -1275$ A1

(2) $\int\left(2x - \frac{4}{\sqrt{x}}\right)dx$

$= \int(2x - 4x^{-\frac{1}{2}})dx$ B1

$= x^2 - 8x^{\frac{1}{2}} + c$ M1A1A1

(3) (a) $\sqrt{48} = \sqrt{16 \times 3} = 4\sqrt{3}$ B1

(b) $(2 - 3\sqrt{3})^2$

$= 4 - 6\sqrt{3} - 6\sqrt{3} + 27$ M1A1

$= 31 - 12\sqrt{3}$ A1

(4) (a) M has coordinates (5, 1). B1

(b) Gradient of $PQ = \frac{6 - -4}{2 - 8} = -\frac{10}{6} = -\frac{5}{3}$ M1A1

Equation is $y - 1 = \frac{3}{5}(x - 5)$ M1A1

$3x - 5y - 10 = 0$ A1

(5) (a) M1A2

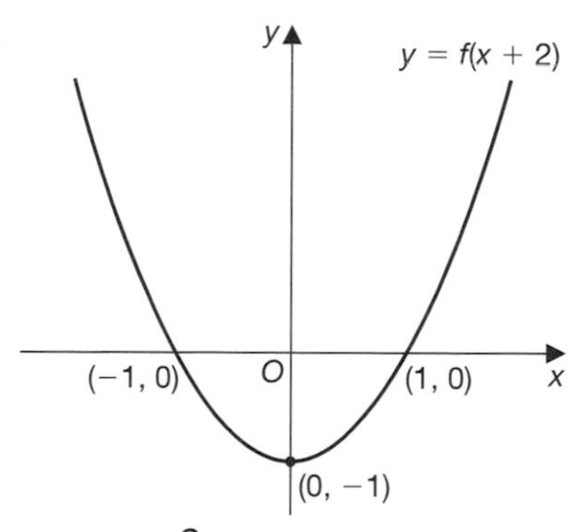

(b) M1A2

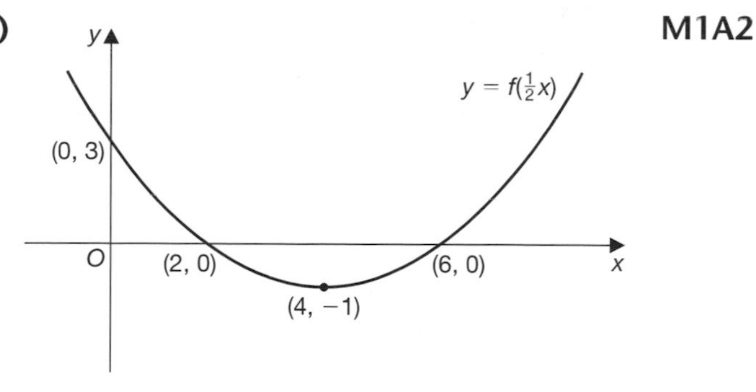

(6) (a) $y = x - 2$ M1

$x^2 + x - 2 - 10 = 0$ M1

$x^2 + x - 12 = 0$ A1

$(x + 4)(x - 3) = 0$ M1

$x = -4$ or $x = 3$ A1

$y = -6$ or $y = 1$ A1

(b) $10 - x^2 > x - 2$

$0 > x^2 + x - 12$ M1

$-4 < x < 3$ M1A1

(7) (a) $6 + 2d = 4y$ M1A1

$d = 2y - 3$ A1

(b) fifth term $= 6 + 4d$ **M1**

$= 6 + 4(2y - 3)$ **M1**

$= 8y - 6$ **A1**

(c) $S_{11} = \frac{11}{2}\{12 + 10(2y - 3)\}$ **M1A1**

$= \frac{11}{2}(20y - 18)$

$= 110y - 99$ **A1**

(8) (a) $x^2 + 8x + 25 = (x + 4)^2 + 9$

$p = 4, q = 9$ **M1A1A1**

(b) $x^2 + 8x + 13 = -12$

$x^2 + 8x + 25 = 0$

$(x + 4)^2 + 9 = 0$

$(x + 4)^2 = -9$

$\therefore$ no real roots **M1A1**

(c) $64 = 4R\ (b^2 = 4ac)$ **M1**

$16 = R$ **A1**

(d)

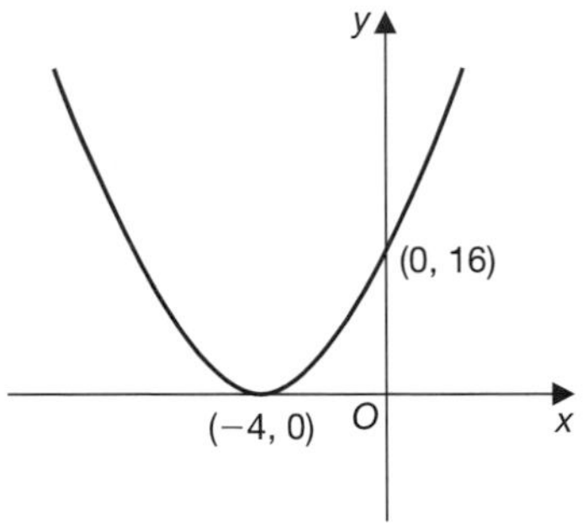

$y = x^2 + 8x + 16$

$= (x + 4)^2$ **M1**

M1A2

(9) (a) $f(x) = \int(3x^2 - 6x + 2)dx$ **M1**

$= x^3 - 3x^2 + 2x + C$ **A1**

$x = 2, y = 3 \Rightarrow 3 = 8 - 12 + 4 + C$

$\Rightarrow C = 3$ **M1**

$\therefore f(x) = x^3 - 3x^2 + 2x + 3$ **A1**

(b) $f(1) = 1^3 - 3.1^2 + 2.1 + 3$

$= 3$ **B1**

(c) $f'(2) = 3.2^2 - 6.2 + 2$ **M1**

$= 2$ **A1**

$3x^2 - 6x + 2 = 2$ **M1**

$3x(x - 2) = 0$ **M1**

$x = 0$ (or $x = 2$) **A1**

(10) (a) $0 = 2 + \frac{3}{x}$ **M1**

$\Rightarrow x = -\frac{3}{2}$, so A is $(-\frac{3}{2}, 0)$. **A1**

(b) $y = 2 + \frac{3}{x} = 2 + 3x^{-1}$ **M1**

$\frac{dy}{dx} = -3x^{-2} = \frac{-3}{x^2}$ **A1**

When $x = -\frac{3}{2}, \frac{dy}{dx} = \frac{-3}{(9/4)} = -\frac{4}{3}$ **A1**

$y - 0 = \frac{3}{4}\left(x + \frac{3}{2}\right)$ **M1A1**

$y = \frac{3x}{4} + \frac{9}{8}$ **M1**

(c) $\frac{3x}{4} + \frac{9}{8} = 2 + \frac{3}{x}$ **M1**

$\Rightarrow \quad 6x^2 + 9x = 16x + 24$

$\Rightarrow \quad 6x^2 - 7x - 24 = 0$ **A1**

$\Rightarrow \quad (2x + 3)(3x - 8) = 0$ **M1**

$\Rightarrow \quad (x = -\frac{3}{2})$ or $x = \frac{8}{3}$ **A1**

$y = 2 + \frac{9}{8} = \frac{25}{8}$ **A1**

AS Mock Exam 2 Answers

Method (M) marks for 'knowing a method and attempting to apply it'.
Accuracy (A) marks can only be awarded if the relevant (M) mark(s) have been earned.
(B) marks are independent of method marks.

(1) $r^2 = (4 - 2)^2 + (-3 - -1)^2$ **M1**

$r^2 = 8$ **A1**

$(x - 2)^2 + (y - -1)^2 = 8$ **M1**

$x^2 + y^2 - 4x + 2y - 3 = 0$ **A1**

(2) $(3 - 2x)^5 = (-2x)^5 + 5(-2x)^4 \times 3^1 + 10(-2x)^3 \times 3^2 + \ldots$ **M1A2**

$= -32x^5 + 240x^4 - 720x^3 + \ldots$ **A1**

(3) (a) $y = \sqrt{20} = 4.472$ **B1**

$y = \sqrt{29} = 5.385$ **B1**

(b) Area $= \frac{1}{2}\{2.236 + 5.385 + 2(2.828 + 3.606 + 4.472)\}$ **M1A2**

$= 14.72$ (2 d.p.) **A1**

(4) $3\sin^2 x - \cos^2 x = 2\cos x + 3$

$3(1 - \cos^2 x) - \cos^2 x = 2\cos x + 3$ **M1**

$0 = 2\cos x(1 + 2\cos x)$ **A1**

$\Rightarrow \quad \cos x = 0$ or $\cos x = -\frac{1}{2}$ **M1A1A1**

$\Rightarrow \quad x = 90°, 270° \quad x = 120°, 240°$ **A2**

(5) (a) Perimeter $= 2(24 \times \frac{\pi}{3} + 48\cos 30°)$ **M1A1**

$= 16\pi + 48\sqrt{3}$ **A1**

(b) Area $= 2\left(\frac{1}{2}.24^2 \times \frac{\pi}{3} + \frac{1}{2}.24^2 \sin 120°\right)$ **M1A1**

$= 192\pi + 288\sqrt{3}$ **A1A1**

(6) (a) Total = £500 $\times 1.008^{12}$ **M1A1**

= £550.17 (nearest penny) **A1**

(b) At end of month 1 $= 200 \times 1.008$

At end of month 2 $= 200 \times 1.008 + 200 \times 1.008^2$

$\therefore$ Required total $= 200(1.008 + 1.008^2 + \ldots + 1.008^{36})$ **M1A2**

$= 200 \times 1.008 \frac{(1.008^{36} - 1)}{(1.008 - 1)}$ **M1A1**

= £8372.19 (nearest penny) **M1A1**

(7) (a) $(2x^3 + px^2 + qx - 18) = (x + 3)(2x + 3)(x - 2)$ **B2**
(by inspection)

$\Rightarrow \quad f(x) = (2x^2 + 9x + 9)(x - 2)$ **M1**

$= 2x^3 + 9x^2 + 9x - 4x^2 - 18x - 18$ **M1**

$= 2x^3 + 5x^2 - 9x - 18$

i.e. $p = 5; q = -9$ **A1A1**

(b) $f(x) = 0$

$\Rightarrow \quad (x + 3)(2x + 3)(x - 2) = 0$ (see part **(a)**) **M1**

$\Rightarrow \quad x = -3$ or $-\frac{3}{2}$ or 2 **A1**

(c) Equation is $f(\sin x) = 0$ **M1**

$\Rightarrow \quad \sin x = -3$ or $-\frac{3}{2}$ or 2 **A1**

$\Rightarrow$ no solutions **A1**

(8) (a) $x + 7 = 10 + 3x - x^2$ **M1**

$\Rightarrow \quad x^2 - 2x - 3 = 0$ **A1**

$\Rightarrow \quad (x - 3)(x + 1) = 0$ **M1**

$\Rightarrow \quad x = 3$ or $x = -1$

$\Rightarrow \quad y = 10$ or $y = 6$

i.e. A is $(-1, 6)$; B is $(3, 10)$ **A1A1**

(b) Area $= \int_{-1}^{3} (10 + 3x - x^2)dx - \int_{-1}^{3} (x + 7)dx$ **M1**

$= \int_{-1}^{3} (10 + 3x - x^2 - x - 7)dx$

$= \int_{-1}^{3} (3 + 2x - x^2)dx$ **A1**

$= [3x + x^2 - \frac{1}{3}x^3]^{3}_{-1}$ **M1A2**

$= 3(3 - -1) + (3^2 - (-1)^2) - \frac{1}{3}(3^3 - (-1)^3)$ **M1**

$= 12 + 8 - \frac{28}{3}$ **A1**

$= \frac{32}{3}(10\frac{2}{3})$ **A1**

(9) (a) $V = \pi x^2 h = 1000$ **B1**

$A = 2\pi x^2 + 2\pi x h$ **B1**

$\Rightarrow \quad A = 2\pi x^2 + 2\pi x \times \left(\frac{1000}{\pi x^2}\right)$ **M1A1**

i.e. $A = 2\pi x^2 + \frac{2000}{x}$ **A1**

(b) $A = 2\pi x^2 + 2000x^{-1}$

$\frac{dA}{dx} = 4\pi x - \frac{2000}{x^2}$ **M1A1**

$\Rightarrow \quad 4\pi x - \frac{2000}{x^2} = 0$ at max/min **M1**

$\Rightarrow \quad x^3 = \frac{500}{\pi}$ **A1**

$\frac{d^2A}{dx^2} = 4\pi + \frac{4000}{x^3}$ **M1A1**

When $x^3 = \frac{500}{\pi}$, $\frac{d^2A}{dx^2} = 4\pi + 8\pi$ **M1**

$= 12\pi > 0$

$\therefore$ minimum **A1**